THE CONSCIENTIOUS GARDENER

THE PUBLISHER GRATEFULLY ACKNOWLEDGES THE GENEROUS SUPPORT OF THE RALPH AND SHIRLEY SHAPIRO ENDOWMENT FUND IN ENVIRONMENTAL STUDIES OF THE UNIVERSITY OF CALIFORNIA PRESS FOUNDATION.

SARAH HAYDEN REICHARD

The Conscientious Gardener

CULTIVATING A GARDEN ETHIC

Foreword by Peter Raven

UNIVERSITY OF CALIFORNIA PRESS BERKELEY LOS ANGELES LONDON

University of California Press, one of the most distinguished university presses in the United States, enriches lives around the world by advancing scholarship in the humanities, social sciences, and natural sciences. Its activities are supported by the UC Press Foundation and by philanthropic contributions from individuals and institutions. For more information, visit www.ucpress.edu.

University of California Press
Berkeley and Los Angeles, California

University of California Press, Ltd.
London, England
First paperback printing 2012

Designer: Nola Burger
Text: 9.75/14 Palatino
Display: Filosofia, Benton Gothic
Compositor: BookMatters, Berkeley
Indexer: Thérèse Shere
Illustrator: Bill Nelson
Printer and binder: Thomson-Shore, Inc.

Library of Congress Cataloging-in-Publication Data

Reichard, Sarah H.
The conscientious gardener : cultivating a garden ethic / Sarah Hayden Reichard ; foreword by Peter Raven.
p. cm.
Includes bibliographical references and index.
ISBN 978-0-520-27275-0 (pbk : alk. paper)
1. Sustainable horticulture. 2. Native plant gardening. 3. Pests—Integrated control. I. Title. II. Title: Cultivating a garden ethic.
SB319.95.R45 2011
635'.048—dc22 2010028833

Manufactured in the United States of America

20 19 18 17 16 15 14 13 12
10 9 8 7 6 5 4 3 2 1

This book is printed on Cascades Enviro 100, a 100% post consumer waste, recycled, de-inked fiber. FSC recycled certified and processed chlorine free. It is acid free, Ecologo certified, and manufactured by BioGas energy.

CONTENTS

FOREWORD

More than two-thirds of the 310 million people who live in the United States actively garden or have an interest in gardening. Americans constitute about 4.5 percent of the world population but consume nearly a quarter of the world's resources. The way in which we live, therefore, has a major effect on the global ecosystem, and our gardening has direct and important effects on our common environment, both locally and farther afield.

In her engaging, personal style, Dr. Sarah Reichard calls attention to the many dimensions of sustainability in gardening at a time when the subject has attained special interest. The "flight to the suburbs" that characterized the United States following the end of World War II led to not only the destruction of a great deal of productive farmland but also the creation of millions of new gardens of a size and scope that had been achieved only by the very wealthy in previous periods of history. In general, the frequency of gardens per individual in the modern, industrialized world is probably as high as it has ever been for any set of urbanized or semiurbanized people in history. The warp and woof of many modern communities largely consists of their gardens and yards. But the network of gardens and the whole style of suburban and urban living in the United States presents special challenges. Flooding from poorly designed gardens erodes stream banks and often causes flooding downstream. Pesticides applied in part of the system drift to other places or contaminate the water running off individual gardens. As the great environmentalist Rachel Carson demonstrated in the 1960s, excessive use of pesticides—including herbicides—not only poisons the environment but soon proves detrimental to human health as well. In gardening as in farming, overdosing with chemicals or disregarding the health of the soil soon leads to ecological disasters. When these disasters are played out in many gardens simultaneously, whole regions suffer.

Dr. Reichard presents the environmental context alongside every strategy for creating a sustainable garden. Particularly timely is the discussion of bioswales and rain gardens. Both capture rainfall and other precipitation near where it falls, and both kinds of gardens are spreading rapidly as people realize the problems associated with the excessive use of hardscape in heavily settled areas—the result is always periodic flooding. (Enlightened utility companies, such as the Metropolitan Sewer District in St. Louis, are basing their fees on charges related to the proportion of hardscape on a particular property instead of on general taxes.) Quite appropriately, gardeners are also increasingly attempting to identify and use native plants, constitutionally well suited to the regions to which they are native. But not all native plants are equal, and some should not be cultivated in home gardens. Many plants have wide ranges and are hugely variable genetically, so everything labeled as a single species is by no means equivalent in its characteristics, and care must be taken in selecting the particular strain of native plant to be cultivated.

Drawing on her rich, practical experience as well as her expertise as an environmental scientist, Dr. Reichard informs us about what we can do to make our gardens more healthy, lively, interesting, and functional. Her research on cultivated plants, especially in relation to their potential to become weeds, her efforts to promote the conservation of endangered native species, and her work with both university students and the general public have equipped her well to explicate the subjects she addresses in this book.

The global human population has tripled during my lifetime. More than one billion of the earth's seven billion people are malnourished, more than one hundred million of them on the brink of starvation. Rates of consumption per person are skyrocketing upward along with population growth, and many of the technologies to which we have become accustomed during the more than two hundred years following the Industrial Revolution are dangerous to human health, the proper functioning of ecosystems, and global sustainability as a whole. Indeed, the Global Footprint Network estimates that we are currently using 160 percent of the sustainable capacity of the Earth on an ongoing basis, up from 70 percent as recently as 1970. There are no easy answers to the dilemma in which we find ourselves, but inaction is not an option. The conscientious gardener fits into the heart of global sustainability and can make important contributions to the health of our planet. This loving and responsible guidebook can direct our steps as we make our way together into the future.

—Peter H. Raven, President, Missouri Botanical Garden, St. Louis

INTRODUCTION

The Land Ethic

> Do we not sing our love for and obligation to the land of the free and the home of the brave? Yes, but just what and whom do we love? Certainly not the soil, which we are sending helter-skelter downriver. Certainly not the waters, which we assume have no function except to turn turbines, float barges, and carry off sewage. Certainly not the plants, of which we exterminate whole communities without batting an eye. Certainly not the animals, of which we have already extirpated many of the largest and most beautiful species. A land ethic of course cannot prevent the alteration, management, and use of these "resources," but it does affirm their right to continued existence, and, at least in spots, their continued existence in a natural state.
>
> —Aldo Leopold, *A Sand County Almanac*, 1949

Aldo Leopold, the biologist, philosopher, and author whose words I have chosen to start this book and whose philosophy guides the chapters, brought together the traditional view that nature should be used for human needs with the more romantic view that nature can refresh and inspire the human spirit. He understood that humans have a place in nature but not the right to exploit it at the expense of all other organisms. He believed that every piece of land is linked to other lands and waters and that action in one place results in reactions in another. Leopold was a professor at the University of Wisconsin and is considered by many to be the father of both wildlife science management and restoration ecology, and his philosophically deep thinking about humans and their relationship to land and all the organisms on it continues to resonate with new generations.

A collection of essays, *A Sand County Almanac* was published in 1949, shortly after Aldo Leopold's death. I first read the book as an undergraduate majoring in botany at the University of Washington—I brought it on a backpacking trip and read it to my boyfriend (now husband), Brian, by flashlight each night in the tent. The final essay, "The Land Ethic," lays out a clear case for changing how humans relate to the land. Our ethics are an expression of our social conscience and direct how we live and interact with one another—ideally to the mutual benefit of all. Leopold encourages us to extend them to include love of the "land," which he defines as a system of soil, water, plants, and human and nonhuman animals. His words struck something deep within me, and over the nearly thirty years since I first read them, I have returned to them repeatedly for inspiration.

Just as Leopold believed that a land ethic reflects an ecological conscience, I believe that a garden ethic reflects the conscientiousness of those who care for land by nurturing gardens. Gardeners revel in the beauty of a flower, the wonders of pollination turning that flower into a lovely or delicious fruit, the snap of a fresh pea pod picked from the vine and eaten on the spot. They are connected to their plot by a love of the living. But the garden ethic also arises from an increasing awareness that, over time, practices and products have crept into our craft that decrease its long-term sustainability. As we have moved from an agrarian society to one based in urban and suburban landscapes, we have lost contact with habits common to our ancestors—such as using naturally decomposing materials rather than synthetic fertilizers to improve soil fertility or nurturing predatory insects and birds instead of deploying the latest, greatest, also usually synthetic products to control pests. A garden ethic gives us the information and structure to return to those less harmful procedures, helping us to view the garden, like the land, as a fully functioning ecosystem—and to incorporate the awareness that its impacts extend far beyond its footprint. Invasive species that escape into nearby wildlands, the mining and transportation of materials such as peat from regions thousands of miles away, and the use of inefficient engines in garden equipment all contribute to the loss of biological diversity beyond our garden gates.

Conservation is important, but as Leopold realized, good intentions can be futile or even dangerous when devoid of critical thinking and understanding. This is certainly true in horticulture today. Many garden practices are based on half-truths, misinformation, and anecdotal rather than scientific evidence. My goal is to provide gardeners with the solid information they need to make meaningful decisions about managing their gardens in ways that respect the interconnectedness of life on this planet, to protect biological diversity across the earth's landscape.

The chapters in this book reflect my evolution as a gardener as well. While I have been a botanical scientist most of my adult life, until several years ago my garden was a tiny patch of earth in a densely urban area. Even with my scientific knowledge, I often failed to see the connections between my actions and their impacts on wildlands. When I moved to a much larger garden, connected to a large greenbelt, I became more conscientious and aware of the consequences of seeming harmless things like raking leaves in the fall rather than allowing them to naturally decompose and feed the soil. I also realized that my previous choices in tending my small garden—even small things like which potting soil I used for container plantings—had repercussions outside my garden.

This book is divided into eight chapters that explore topics important to both gardens and conservation. Fertile, porous soil and clean water are critical to our existence, but garden practices affect their health and sustainability; these natural resources are explored in the first two chapters. Chapters 3 and 4 guide plant selection. What are native plants, and should you use them? When are they appropriate, and when is a nonnative a better choice? How can you determine which nonnative species will invade, and why are people so concerned about them in gardens when the problems they cause occur in wildlands? The plants you select, as well as other aspects of your garden, such as its structure and water features, can attract desirable wildlife and repel undesirable animals, topics explored in chapter 5. However, gardens also invite unwanted plant, insect, and other species, and how to safely control and even prevent their presence is the subject of chapter 6. The final two chapters, on global warming and reducing waste, look at the big picture as a spur to change. Gardeners can help prevent climate change through simple measures such as reducing soil tillage, switching to push mowers, growing some of our own food, and planting trees to shade the house in summer. We can also help shrink landfills by composting, not buying overpackaged goods, and either reusing or freecycling garden items.

The book ends with an appendix summarizing more than twenty years of my research on garden plants that can become invasive, with a table of such species worldwide, their impacts, and where they are known to be problems. It is followed by a glossary and a two-part resources list for each chapter: References for Gardeners has publications with more information for those wanting to explore specific issues, while Technical Papers lists scientific studies with documentation and deeper explanations.

We often expect government—local, state, and federal—to regulate to promote the conservation and protection of natural resources, and this is often appropriate and necessary. But even in the absence of such laws, the garden ethic, like Leopold's land ethic, reminds us of our obligation to remember that we are part of a community that includes plants, animals, and a myriad of other organisms that are integrated with and dependent on one another. The garden ethic also reflects conscientious choices about how we treat this community, for both its good and our own. Developing your garden ethic will take some effort, but it will be a journey of discovery resulting in greater satisfaction with your garden—and maybe yourself.

CHAPTER ONE

The Skin of the Earth

Land, then, is not merely soil; it is a fountain of energy flowing through a circuit of soils, plants, and animals.

—Aldo Leopold, *A Sand County Almanac*

We all have a relationship with soil. Some may call it "dirt" and try to launder it out of their children's clothing. Some may use it to grow crops to support their families, some may try to stop it from washing downhill in storms, and some spend their lives studying it and mapping it. Many gardeners think of it mostly when they are removing it from under their fingernails after a day planting bulbs, but anyone who wants to grow plants should know something about it, and surprisingly few do.

When I was a botany student I was not required to know anything about soil science—I learned a bit about soil formation in an ecology class, and that was it. Even today, many professional botanists and even horticulturists know surprisingly little about soil. But think about it: a substantial amount of the biological mass of a plant is in the roots, which need soil to support and nourish the parts aboveground.

Soils are incredibly dynamic—complete ecosystems that are constantly, though sometimes slowly, changing. They have mineral components but are teeming with life, such as fungi, beetles, worms, and gophers. Under most natural conditions they form at a rate nearly equal to that of their erosion, which happens naturally by means of water, wind, and gravity. They are neither permanent nor indestructible.

Skin Deep: Soil Formation and Composition

As soils develop, they usually form a profile of different layers (see figure 1). On top, there might be a relatively thin layer of organic material made up of living, decaying, and decomposed plants and animals. This "O" horizon may be only a few inches thick and is often darker than lower layers. Below that is what soil scientists call the "A" horizon, composed of fine mineral particles (sand, clay, and other inorganic matter) with some accumulated decomposed organic material. Moving deeper are "E," "B," and other horizons, mostly mineral particles of increasing size. Eventually, the soil stops at hard bedrock.

These layers, along with rocks, combine to create what has been appropriately called "the skin of the earth." Like skin, it is rich in variation—of color, moisture, graininess, and so on—and its multitude of types can be classified by these and other characteristics. The United States alone has more than twenty thousand identified soils!

Scientists who study soil formation consider five interrelated factors. The first is the parent material. This usually comprises large igneous rocks, created from volcanic activity; sediments that become solid through intense pressure; or combinations of the two into metamorphic rocks through heat and pressure. The rocks are slowly weathered by physical, biological, and chemical processes to form smaller and smaller particles. Then wind, water, glaciers, and gravity may move these particles long distances, so parent material in a location may not be the only or even the principal component of the soil found there.

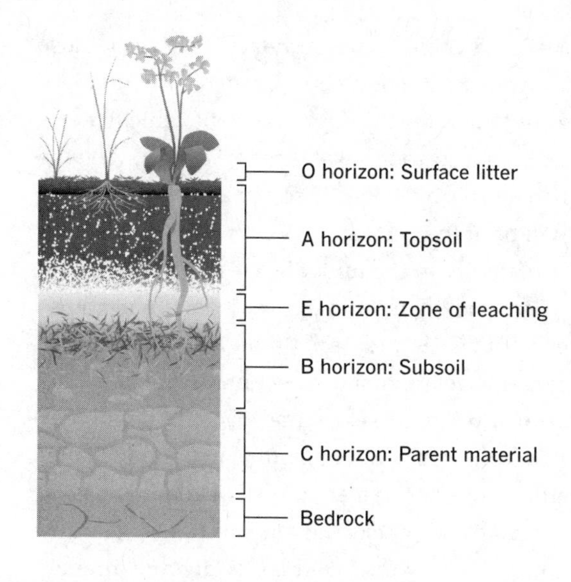

Figure 1. A typical soil profile showing the common layers. Some layers may be missing in some soils, and the E horizon, which contains fine particles that wash out of A, may be very small. Used with permission from SparkNotes LLC.

The second factor is the climate, primarily precipitation and minimum and maximum temperatures. Wetter climates usually foster more plant growth, which means more organic matter to break up rocks (such as tree roots pushing through concrete) and then decaying to become part of the soil. Water also plays a direct role in weathering parent material. Areas with high precipitation generally have thicker soil layers, especially if the rain falls when the soil is warm, leading to increased chemical weathering as organic matter decomposes and releases organic acids. In colder weather, water can get into crevices in rocks and freeze—exerting a force of 150 tons of pressure per foot and breaking the rocks into smaller pieces—then thaw, releasing the fragments and allowing water to reach deeper into the rock.

Topography is the third factor. Aspect, or the relation of a place to sun and weather, plays a big role in the temperature of a site. In the Northern

Hemisphere, north-facing slopes are cooler than south-facing slopes because they get less direct sunlight, while in the Southern Hemisphere the opposite is true. Another slope consideration is the degree of slant, which affects the amount of erosion that will occur through water runoff and gravity: the steeper the slope, the less distinct the layers of the soil profile will be at the top and the wetter the soil will be at the bottom.

The fourth factor is biotic material. In the upper layers of soil there may be up to five tons of living organisms (roots, animals, bacteria, fungi, and so on) per acre. On the surface, plant debris and dead animals decay and provide nutrients, and aboveground vegetation may slow rain as it falls, reducing erosion. Belowground, roots, worms, beetles, and bacteria help break down the decaying material, adding substantial amounts of organic matter to the soil. This organic matter contains large amounts of carbon, which is sequestered underground instead of being released into the atmosphere and contributing to global warming. Roots, worms, and larger digging animals help mix these soil layers and aerate the earth.

The fifth and final critical factor is time. Soils of all ages exist around the world, each in its own stage in the cycle of creation and destruction. As time moves on, all the other factors come into play: large rocks form and erode, local organic matter increases or decreases, and weather takes its toll.

Keeping the Pores Open

Perhaps the most important consideration in soil health is having sufficient pore space between particles to allow water and air to reach plant roots. (Some guidelines suggest that the optimal soil for plant growth has only 5–10 percent organic matter, 40–45 percent mineral particles, and about 25 percent each of water and air.) Roots have important jobs to do, including anchoring the plant, finding and absorbing water and nutrients, storing carbohydrates for later use, and connecting with helpful fungi (described below). Woody roots absorb some water and nutrients, but much of the root's work is done by its more tender ends and hairs, which form on its surface. The roots move through pore space and through larger openings created by earthworms and other soil fauna. If the soil is very compacted, by such things as construction equipment or heavy foot traffic, the roots will not be able to move and capture enough resources or to produce a wide or deep base, affecting their ability to stabilize the plant. Soil compaction is both bad for plant health and difficult to undo: it leads to less animal activity, which means less new pore space creation. Adding a thick layer of wood chip mulch to high-impact areas may reduce compaction.

One measure of compaction is tilth, or the composition of aggregated clusters of mineral grains. The word shares its root with the verb *till,* to loosen soil manually. Tilling can happen naturally, as through frost heaving or worm or insect movement. Under most circumstances, deep manual tilling is unnecessary and may even destroy soil structure, leading to a long-term loss of pores and the chopping up of integral fungi and soil invertebrates. Deep tilling should only be done for removing lawns or preparing beds for root crops, in soil that is at least 60°F and not too wet. If the soil remains in a ball instead of breaking up when you squeeze it in a fist, give it time to dry.

The Living Soil

Healthy soil is teeming with life. The smallest organisms are the many bacteria that, along with worms and insects, are critical to breaking down organic matter and converting soil nitrogen to forms usable by plants. Larger animals, such as moles, gophers, rabbits, and mountain beavers, live underground in what can become vast burrow systems, creating large, porous spaces. Fungi provide many valuable services, and, of course, there are all those plant roots. Each garden has a veritable city belowground!

The soil biota may affect human health. A 2007 study by Christopher Lowry and Graham Rook suggests a reason why gardening feels so good:

a bacterium naturally found in soil, *Mycobacterium vaccae,* stimulates the human immune system to release serotonin. This hormone is used in antidepressants to increase feelings of well-being. Some scientists even believe that our ever-increasing desire for cleanliness and our distance from farming activities are leading to health problems such as asthma and allergies. Perhaps doctors someday will prescribe gardening for a healthy life.

Cultivating the Fungus among Us

Fungi are found not just in the threadlike networks that hold soils together but in plants themselves. They often attach to the roots, and studies such as those by Regina Redman and her colleagues have found them in stems and leaves, where they increase the plant's cold and salt tolerances. They commonly take the form of arbuscular mycorrhizae (AM), fungal threads that grow branching structures inside the root, where they receive carbohydrates in exchange for protecting the plant against disease and helping it to absorb water and nutrients.

Very little is known about AM fungi in most ecosystems, but many things are implicated in reducing their amount or type: extensive impervious surfaces, excessive disturbances and tilling, topsoil removal, and the use of pesticides and fertilizers. New landscapes have been found to have fewer AM fungi communities than established ones, probably because of the recent soil disturbance during construction. Too much phosphorus (the middle number on the fertilizer package) can also cause problems, by decreasing the ability of AM fungi to colonize plant roots, making them work harder to extract water and nutrients from the soil. Robert Linderman and E. Anne Davis have found that fertilizers in general, even those with less phosphorus, lead to less root colonization.

Mycorrhizal inoculation kits are commercially available to remedy these losses, but smart gardeners will avoid them. These kits can be ineffective—not all plant families use AM fungi—or even cause harm: nonnative AM fungi may change the community structure of native fungi and may replace fungi that local plant species rely on. If you focus on the overall health of your soil and use phosphorus-containing products in moderation, appropriate fungi will find your garden.

I Feel the Earth Move

Time, weather, and the incredible heat and pressure that form large rocks are implacable forces, but their effects on soil are increasingly outstripped by those of living organisms. Humans are perhaps the prime example: our farming procedures can rapidly deplete the soil of nutrients, land clearing can contribute to erosion, and construction practices can compact soils, making them less suitable for plant growth. We think nothing of moving soil to clear foundations for new buildings or even to reshape local topography. The city I live in, Seattle, was built on several hills. In the decades shortly before and after 1900, to flatten out and expand the town, the city leaders moved about fifty million tons of soil from some of the hills into the harbor and other areas! These neighborhoods are still often referred to as the Regrades.

But humans aren't the only large-scale earth movers. Worms turn too. Charles Darwin was fascinated by worms and spent decades studying them. He estimated that they brought more than ten tons of earth to the surface each year on every acre of English land. You might see this process in your own garden, as I have. Late in the winter of the first year I lived in my new house, I was startled to find small mounds of leaves dotted across the ground in my woodland garden. I pushed several aside and in every case found a small hole, each of which held an earthworm. After grabbing the leaves with their mouths and pulling them down the hole, the worms consumed them and defecated the remains, mixing organic layers into the inorganic.

It turned out that my earthworms are actually a widespread European species, *Lumbricus terrestris,* the very same worms that Darwin studied. Nonnative worms are found all over the world except at the ice caps and in arid lands. A Brazilian worm, *Pontoscolex corethrurus,* has spread through the tropics, several destructive Asian species in the genus *Amynthus* have invaded the eastern United States and are moving westward, and the species of worm in my garden—the common nightcrawler known to fishermen and dissected in my seventh grade biology class—has colonized many temperate areas. (Ironically, it is becoming rare in Europe, where two introduced flatworms prey upon it.)

Where native worms already exist, new introductions like these may overconsume local food sources. But perhaps the most serious problems occur where there are no native worms, such as the northern temperate forests of North America, which have not had native worms since before the last glaciations thousands of years ago, as shown in figure 2. There, the worms modify the soil structure, affecting the flora and fauna. Studies in hardwood forests by Cindy Hale and her colleagues at the University of Minnesota show a steep decrease in plant abundance and species diversity along a leading edge of earthworm invasion as the number of earthworms increased. Populations of a rare fern disappeared. The worms eat seeds, which may play a role in decreasing biodiversity, but they do the most damage in greatly altering the forest litter layer, exposing seeds and spores to predation and the desiccation that results from the disturbance of the moist, nutritious soils needed for germination.

Although nonnative worms are responsible for many problems, don't worry too much about the ones you already have. Just follow these few tips to keep them contained. Dispose of worms found in nursery material and never use soil with worms as fill dirt, especially if you live near a wooded area. Worm bins, popular for helping turn food waste into compost, should never be dumped anywhere. Freezing the compost for one to four weeks will kill the worms, but better yet, do not use worms at all: let naturally occurring organisms break down the waste. If you do buy worms to assist

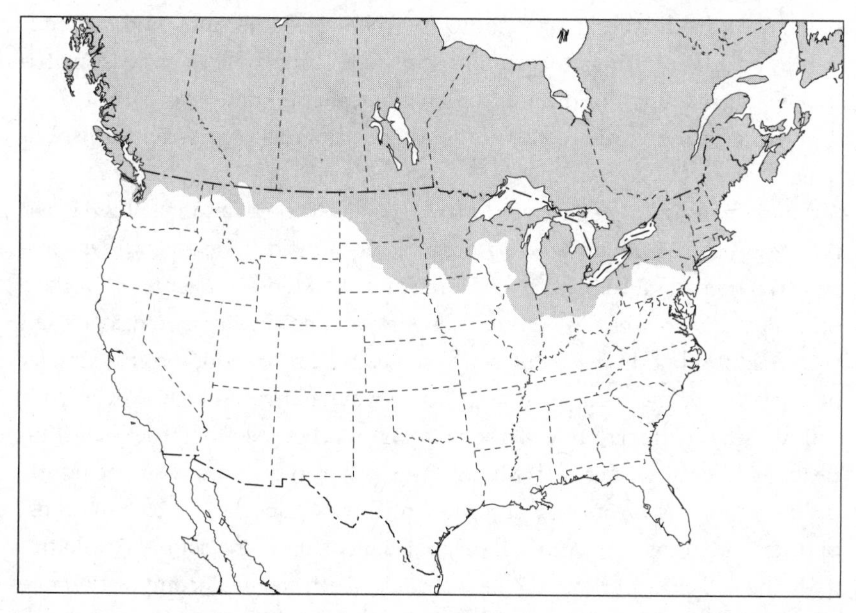

Figure 2. Glaciers covered the northern part of North America until 11,000 to 14,000 years ago, and in places the ground to their south was frozen for hundreds of miles. Earthworms are not native anywhere the ground was frozen at that time, represented by shading on this map. Adapted from original by University of Minnesota, Cindy Hale.

in composting, ask your supplier the name of the species and learn more about it—but know that earthworm identification is difficult and species may be sold under incorrect names. The University of Minnesota website listed under Resources can help you learn more.

Using Soil Amendments

Many of us are aware of the importance of healthy soil for healthy plants and are concerned that our soil is inadequate. If it is unable to grow certain plants, we are told that we should improve it—sometimes with amendments, which are things incorporated into the soil profile to increase nutrition, water retention, or drainage. Soil amendments come in organic (plant-derived) and inorganic (generally made of minerals or plastic) forms, but none is perfect.

Before considering soil amendments, you should get an idea of what soil type you have. Avid gardeners would ideally consider the soil at their prospective home before buying it. Libraries and extension agents often have soil maps of the area, so you can see how soil scientists have categorized your neighborhood. These maps are increasingly also found online, usually on local government pages. If your home was built recently, however, your soil type may differ from what the map says it is because construction activities may have altered it with fill dirt from other locations or compacted it.

You might also want to test for key nutrients, especially nitrogen, phosphorus, and potassium, and soil pH (pH, a measure of the acidity or alkalinity of the soil, is important to plants because it affects the decomposition rate of many biotic components and plants' ability to take up various nutrients). Soil that is healthy for most plants has a neutral pH, though many species prefer soil that is slightly acidic or alkaline. Nitrogen is important for foliage growth, phosphorus increases blooming, and potassium encourages healthy roots. These are the three numbers commonly found on fertilizer packages, in this order. You can purchase soil testing kits at garden supply stores, but you can get better results from professional tests for a relatively nominal fee. Check with extension programs to identify some of the companies and organizations that offer this service (such as the University of Massachusetts Amherst, whose soil-test web page is listed under Resources).

Once you understand the fertility and texture of your soil you are in a position to plan your garden. If you want to grow rhododendrons in heavy, alkaline soil, your first impulse might be to rush out to buy amendments to create a more acidic soil. But there are several things you should consider first.

There is a limit to how much you can do to improve your soil. The "skin of the earth" is complex, and manipulating it is usually about as easy and successful as permanently changing your own skin. Digging in amendments to a depth sufficient for root growth, for example, is expensive and difficult to accomplish on a garden-size scale, especially to the rooting depths of trees and woody plants, such as rhododendrons or oak trees, which have root systems that grow to from two to thirty feet deep and perhaps ten feet wide, depending on the species. Thoroughly mixing the amendments with the native soil is also challenging, and if you do not combine them thoroughly, you risk creating an interface of different soil textures. Roots often do not move easily through soils with different textures, which limits their growth, rendering them unable to fully support the plant or adequately forage for water and nutrients. Also, as previously noted, you run the risk of destroying pore space if you add the amendments when the soil is wet.

To avoid the temptation of soil amendments and fertilizers, garden within the normal ranges of pH and organic matter of your native soil type. Work with your soil, not against it: when planting containerized plants, break up the root ball, removing as much of the potted soil as possible, spread the roots, and backfill the hole with the soil that came out of it, top-dressing with an organic mulch. The mulch will slowly decompose, adding all the nutrients the plant should need.

One of the most common soils people want to amend is heavy clay with poor drainage and aeration. To improve it, 25–50 percent of the soil volume must be amended. It may be possible to do this to the rooting level of perennials in a small area, but not to that of woody plants. If you are determined to amend clay, use only composted leaves and pea gravel. Adding sand will only make the situation worse.

If you are growing herbaceous plants or container gardening, it is all right to use amendments. Just remember that you generally need only 5–10 percent organic matter, so don't overdo it. If you're making a perennial or vegetable bed, dig amendments in at least a foot deep, and ideally deeper.

Many vegetable gardeners prefer creating a raised bed and filling it with purchased topsoil.

In most situations, amendments only introduce nutrients that quickly leach out of the soil. Their collection and production can also harm the environment, so if you must use them, select sustainable materials. In the following pages I discuss several organic and inorganic options.

Organic Amendments

Commonly used organic amendments include straw, grass clippings, compost, manure, wood chips, sawdust, and peat *(Sphagnum)* moss. The most highly recommended amendments are sustainable, meaning that they can be replaced at a rate at least equal to that of their removal. For instance, straw and grass clippings are produced by the removal of aboveground plant parts. The roots can then regenerate the aboveground growth, or the plants and more seeds can be reseeded into the harvest area for yet another crop. Compost is decomposed weeds or clippings from plants that would otherwise be thrown out. As for manure . . . well, there is no danger of it being produced in insufficient amounts. Just make sure that the manure you use—whether steer, chicken, horse, or zoo animal—is matured to eliminate ammonia and other harmful chemicals that can chemically burn your plants. Sawdust and wood chips, added to increase the permeability of soils to air and water, are usually by-products of timber milling. With the rise of sustainable forestry practices, these products are also increasingly sustainable. However, some horticulturists think that bark and sawdust, especially as mulch, do not improve aeration very much and may even adversely affect it by forming dense surface layers. Wood chips, however, do not form those layers. The use of peat moss is hotly debated and is discussed below.

Compost and manure, in particular, are used to increase soil fertility because they are high in nitrogen. (Biosolids, or treated sewage sludge, are not appropriate for home use.) This element is an important nutrient for plant growth, and it is sometimes necessary to add more to the soil, especially if the soil doesn't get enough from fallen leaves. (The microorganisms that decompose organic matter also require nitrogen—but the amount deposited by precipitation is usually enough.) Before you add nitrogen, however, test to determine if your soil needs more. If it gets too much, some might end up in streams and lakes (see the following chapter). And more nitrogen might mean more pests: vast numbers of studies, including a 1984 paper from the noted entomologist Thomas White, have shown that plant-eating insects, which also need nitrogen to grow, are attracted to plants with higher nitrogen levels. Compost or manure will also be more effective used as a mulch than as an amendment, because the nitrogen will leach more slowly into the soil when it is irrigated.

***Sphagnum* Peat Moss** If you have indoor or outdoor potted plants, chances are good you use a commercial potting soil mix. And if you do, chances are good it has peat: so-called soil-less mixes rely heavily on this material. In addition to peat, potting mixes usually contain compost, soil, and sand—a general mix has about equal parts of the three key ingredients, though specialty mixes often vary the proportions—plus other ingredients, perhaps including blood and bone meal, for added nutrients. (If you would like to make your own, see the Organic Gardening website under Resources and remember to use one of the recommended alternatives to peat.) Peat is popular with potting soil companies because it holds moisture well, meaning gardeners have to water less. It also is good for starting seeds because it seems to foster less of the damping-off fungi that plague seedlings, especially when they are warm and wet. Its acidity is also good for growth in many plants. However, there are several good alternatives to peat, summarized in table 1, and many reasons not to use it. Suggest to the management at your local garden center that they provide mixes with peat alternatives.

Peat is produced worldwide in wetlands that are important for a number of reasons (as described in the following chapter). Peat bogs exist in 180 countries, appearing on every continent but Antarctica, in both tropical and temperate areas—especially colder temperate areas. They cover at least 2.5 million square miles in total. The Convention on Biological Diversity, an international treaty with 168 signing nations, has noted that it is important to minimize the degradation and to promote the restoration of peat bogs. In 2002 the Ramsar Convention on Wetlands, developed by an international group of scientists in 1971 to protect such regions and now with 159 signing nations, established a committee to monitor progress on the implementation

TABLE 1. ALTERNATIVES TO *SPHAGNUM* PEAT

Material	What Is It?
Coco peat, coir, coir dust	Coco peat and coir are made from the fiber found between the husk and the outer shell of a coconut. Coir dust is a by-product in the manufacture of brushes, mats, and other items.
Compost	Decomposing organic matter, which may include leaves, grass, manure, bark, and nutshells
Reed or sedge peat (sometimes called Michigan or Florida peat)	A peat product made from reeds and sedges and other wetland plants
Dried alfalfa	Alfalfa that has been dried, ground, pressed through a screen, mixed with water, and then dried again for about forty days

of its Guidelines for Global Action on Peatlands. There are meetings, which include scientists as well as industry representatives, every four years to assess the status of peat across the globe.

Peat bogs are dominated by *Sphagnum* moss, whose many species are all found in wetlands. *Sphagnum* has an interesting growth pattern: its stems and strands may be quite long—but they are alive only in the top inch or so. The lower, dead part is shaded and often submerged in water. Because peat favors waterlogged, acidic, and low-oxygen conditions, it grows very slowly, but the mats of dead vegetation may be one to several yards thick. The lowest parts of the stems may have been in place for centuries!

Characteristics	**Sustainability**
• Similar or better color, water absorption, drainage, cation exchange capacity (a measure of fertility) • Supports few weeds or pathogens • Higher in sodium and potassium—may help growth • Higher pH, but fine for acid-loving plants • Some sources have high salinity—buy coir labeled for horticultural use	• Coconut trees can grow in salty soils where little else grows, and they produce coconuts continuously. • Coir dust is a waste product, and using it helps solve the problem of disposal.
• Characteristics vary by source material (e.g., conifers, herbaceous plants) • Similar water absorption • Generally higher in nutrients, especially nitrogen	• Compost is completely sustainable and keeps the materials it uses from going into landfills.
• May have higher nutrient content • pH may vary from 4 to 7.5, depending on source material • Does not hold water as well, but improves water retention in containers	• Requires the dredging of wetlands
• Not as good a growing medium • Higher in nutrients • Holds water well	• Can be farmed sustainably, but it is not widely available

Much of the current mining of peat is for horticulture, but in some parts of the world, such as Ireland, it is an important traditional fuel source still in use, and in Scotland distillers use peat to give scotch its lovely smoky flavor. The process of obtaining peat moss can be quite lengthy. Ditches are dug to drain 95 percent of the water from the bog, which can take a year or more. The peat is then cut or vacuumed out, destroying bogs that may have been developing for centuries, and further dried and ground before being packaged and sold.

There are many reasons to preserve peat bogs. Some bogs may be thousands of years old, and they are important records of human and plant history because their acidic, low-oxygen conditions are excellent for preserving biological materials. In 1967, Canadian botanists were able to germinate ten-thousand-year-old lupine seeds found in a Canadian peat bog! In a more gruesome case, in 1950 two Danish men found a corpse while cutting peat to burn for fuel. Thinking they had found a murder victim because of a noose around its neck, they notified the police, who, baffled by the case, turned to an archaeology professor. The so-called Tollund Man (named after the village where he was found) was a murder victim, all right. However, he was two thousand years old, so the perpetrator was long dead. The typical peat bog conditions had preserved his soft tissues so perfectly that the features of his face, including wrinkles on his forehead and stubbly beard on his chin, were clearly distinguishable.

Peat bogs are also home to many living creatures. While bogs around the world have many things in common, such as species of *Sphagnum* and acidic, low-oxygen conditions, each has a unique collection of biota: the communities of plants and animals that live in bogs are not the same everywhere. Bog biota can also be unusual. For example, carnivorous plants, those curiosities of the plant world, are frequently found in peat bogs. They have evolved the ability to trap insects (despite the exciting name, they rarely snare larger animals) because peat bogs are so low in other sources of nutrients. These plants trap their prey with sticky hairs, pitchers of poison, or leaves that snap shut before the plant chemically dissolves the decaying bodies, absorbing their nutrients. Such plants usually have very limited ranges. The decrease in peat bogs is likely to harm them and some of the other species that live in these isolated, specialized habitats.

Humans depend on peat bogs too—even from afar. In recent years peat bogs have been recognized as a huge sink, or repository, of carbon. It has been estimated that northern latitude peat bogs hold 455 billion metric tons of carbon, which is slightly less than the amount estimated to be in all other living organisms. There is mounting evidence that large amounts of atmospheric carbon dioxide (CO_2), released through the burning of fossil fuels, lead to increased temperatures worldwide and have consequences in every part of this planet. The peat bogs' capacity to capture and hold carbon is critical, and there is debate about how much impact mining them will have on climate change. Defenders of peat mining, for instance, claim that the huge expanses in areas too remote to mine still sequester millions of tons of carbon each year. They also claim that peat bogs produce methane, a greenhouse gas like CO_2, which would mean that active, or growing, peat bogs are positive contributors to global warming. Environmentalists counter that CO_2 emissions are increased by draining the bogs prior to mining, which increases decomposition of peat and other organic components. Most independent scientists agree with environmentalists that the carbon-holding capabilities of peat bogs are important in preventing climate change, and they further say that the bogs preferred for mining produce relatively little methane. The Intergovernmental Panel on Climate Change has determined that when peat bogs are destroyed, the harm from increases in other greenhouse gases, such as carbon dioxide and nitrous oxide, far outweighs the benefits from the decrease in methane. (See chapter 7 for more on climate change.)

Why can't we just regrow the bogs after a harvest? This happens to some extent, but because *Sphagnum* grows so slowly, abandoned peat bog communities do not grow back quickly, even when the ditches are removed and the water system is restored, and peat is not replenished at anywhere near the rate at which it is mined. Susan Jonsson-Ninniss and a colleague found that although a peat mine had unassisted growth soon after it was abandoned, after two decades its community had only about 51 percent of the species found in the undisturbed parts of the bog. Perhaps more disturbing, researchers noted that the regenerated area had significantly more trees than the undisturbed parts, especially birches, suggesting that mining peat from bogs fundamentally alters their structure.

Can we do anything to help the bogs grow back? Ecological restoration works to help disturbed ecosystems return to their original state more quickly. Line Rochefort and colleagues collected a small amount of four *Sphagnum* species from undisturbed bogs. They sowed both live whole strands and shredded bits across exposed dead peat and found that the size of the moss fragments did not affect peat regeneration. A layer of no more than about three-quarters of an inch was most effective. If it was thinner, they grew mostly *Sphagnum;* if thicker, they got mostly vascular plants. The restoration sites with a protective mulch of straw and some additional phosphorus also grew better. The study proved that bogs can grow faster than by natural regeneration, thus increasing the sustainability of peat mining. However, these experiments lasted just five years, so we have little information about the effects of such restoration efforts over time.

Humus Humus from Alaska is promoted as a source of compost and is sold in parts of the United States. Like peat, it takes thousands of years to form. While it may be necessary to remove some in areas where there is building construction, an increase in demand will increase the probability that it is mined (like peat) in nonconstruction areas, so use other forms of sustainably and locally produced compost instead.

Biochar Sometimes an idea catches on like wildfire before much thought or research supports it. So it is with biochar, a fancy charcoal that is sweeping the agricultural and horticultural worlds with amazing speed. It is essentially charcoal that is made by burning wood at a low temperature for a long period. As wood burns, some of its carbon compounds become gases and some remain as solids, such as cellulose and lignin. Biochar's slow burn means about 25–35 percent of the original dry mass remains. According to Professor Sally Brown at the University of Washington, however, biochar is less effective as a soil amendment than compost because it does not attract soil microbes, which improve soil structure and water retention and help the soil hold important nutrients such as nitrogen and phosphorus. Char *is* effective at holding carbon in the soil, but Dr. Brown points out that char does not really help plants grow, and compost excels at that—and fast-growing plants sequester carbon very well!

Inorganic Amendments

Inorganic amendments are used to alter soil pH and to change soil structure. As with organic amendments, you should look for products from sustainable, environmentally safe sources. Wood ash is sometimes used for acidification, but its content and pH vary, and its use is not advised. Ground limestone makes soil less acidic while adding calcium and magnesium. Limestone quarries may or may not be friendly to the environment, mostly depending on how local political agencies regulate them, so you may wish to avoid using limestone unless you can investigate the source. Finely crushed oyster shells might be a better solution: like limestone, they can be applied to increase the soil pH and are a rich source of calcium and magnesium, and they are sustainable as long as the shellfish are harvested sustainably.

A few other amendments are used primarily to improve soil structure. Gravel, perlite (a silica mineral), and sand are quarried or mined and incorporated to increase drainage. The same caution about quarries applies. Note that the sand should be of low salinity to avoid injury to plants—look for horticultural sand.

Some gardeners attempt to improve the water-holding capacity of their soil, especially in containers, with vermiculite. Vermiculite is a naturally occurring mineral that is mined, then heated to extreme temperatures, which makes it swell. Vermiculite has been used for many purposes and is often included in soil mixes. Most of the vermiculite mines that have been tested have deposits of asbestos, fibrous mineral crystals that can cause severe lung damage and lung cancer. Unfortunately, it doesn't stay in the mines. The U.S. Environmental Protection Agency tested sixteen commercially available potting mixes and found detectable levels of asbestos in five, while one released airborne fibers with simulated use in a lab. Consumers have no way of knowing which bags contain asbestos, so the EPA recommends that buyers use the products outdoors, thoroughly wet them (to prevent airborne fibers), and be careful about bringing clothes indoors afterward.

It's easy enough to avoid these risks: forgo the use of vermiculite and instead grow plants suited to the precipitation of your climate, employing mulches and good soil practices to improve soil moisture.

Using Mulch

Any substance that is laid on top of the ground can be considered mulch: bark, wood chips, rocks, recycled glass, cardboard, newspaper. Even some plants and ground covers are used as living, or green, mulch. Each type of mulch has unique properties and production, with implications for conservation. Like amendments, mulches can be organic or inorganic. Organic mulches may provide some nutrition to the soil, while inorganic ones may increase soil temperatures and are usually long-lasting. Given that there are practically an infinite number of things that can be used as mulch, with new ones introduced all the time, I will cover only the most common types, summarized in table 2 on pages 28–29. Use mulches that are produced locally to reduce their carbon footprint (see the chapter on climate change for more information).

Unlike with soil amendments, there are many good reasons to use mulch. One of the most common is to increase the retention of moisture. Exposed soil is subject to heat and wind, which increase evaporation. A protective layer of mulch can shield water from their effects. Most mulches will hold some water, making precipitation a little slower to reach the soil, but the retention benefits result in an overall increase in moisture. In general, organic mulches retain soil moisture better, with the exception of gravel, which is also effective. Not surprisingly, living mulches do not increase soil moisture, because they use the water themselves (although some, such as kinnikinnick and creeping rosemary, are efficient water users).

Mulch can decrease the germination of weeds, which also compete with landscape plants for available water. Weeds can increase the loss of water from the soil by up to 25 percent. They may also steal sunlight if they cover or are higher than your landscape plants. Existing weeds may be stressed by a mulch covering, but it is most effective to remove them *before* applying mulch. I have managed to tame amazing infestations by weeding an area and then directly applying at least three inches of bark or wood mulch.

Many people use mulch for temperature control. In hotter regions, organic mulches can lower soil temperature as much as 50°F! They can raise the temperature, too: a protective layer can prevent roots from freezing. Just be sure not to bunch the mulch around the stem, because a lack of air circulation and trapped moisture can lead to rot. Inorganic mulches, such as stones, store and conduct heat and are sometimes used to increase the temperature around heat-loving plants. Mulches can also help the soil heat up earlier than bare ground in the spring, giving plants an early boost, and plastic sheet mulches, especially clear plastic, can kill weeds by raising the soil temperature, a process called solarization.

According to Dr. Linda Chalker-Scott, known to many as the Mulch Queen and the author of a definitive review of mulches, the word *mulch* is derived from the Germanic term *molsh,* which means "soft" or "spongy." It is therefore not surprising to learn that mulches are a great way to prevent soil compaction. Just a small layer of mulch can help reduce the impact of feet, equipment, vehicles, and even pounding rain. (If you are using mulch to protect tree roots, lay it as far as the branches extend.) Mulch can help prevent surface-soil erosion by cushioning the impact of rainfall. However, living or any other mulch cannot stabilize slopes—this requires more deeply rooted plants such as trees.

Organic Mulches

The most common organic mulch is probably bark chips, from conifers or hardwood species. It is a by-product of timber milling—the bark is stripped off the felled tree, then chopped up and sold in bags or in bulk. Because this is waste material, as long as the trees are harvested in a sustainable way, the product is fully sustainable. Some barks, such as from fir trees, decompose quickly, requiring more frequent application, and many will give you terrible splinters, so wear gloves when working with them. Bark tissue is waxy, so the chips might not absorb and rerelease water into the soil. Some bark mulch may also come from trees held in salt water before milling and may be very saline.

Related to bark are wood chips. They are sometimes made from recycled wood products, such as old pallets, or are a side product of industry. In areas where paper is produced, some trees are chipped to produce pulp. The larger chunks may be sold as mulch or for playground surfaces. If you purchase bulk wood chips, which is how they are usually sold, from a supplier, you should ask where they are from and how the trees are harvested. A bulk supplier likely buys directly from the mill and should know or can ask.

In most places wood chips do not result in a substantial loss of forest trees. In fact, mulch from urban arborists, made from tree and other plant trimmings, would end up in landfills if not used. Many arborists, looking for a place to dump their chips without paying a fee, will bring them to your house for free or at a low cost. (Another free source of mulch: if you need to remove a tree from your property, have it chipped and left in place.) Arborist mulch is also good because it usually includes pieces of varying sizes, which prevents it from becoming packed down to form a barely penetrable layer, something that occasionally happens with bark. Be sure, however, that the source of the wood is local: some plant pests have been transported in wood products into previously uninfected areas.

There are, however, places where wood chips are harvested unsustainably. In the southeastern United States, forests of bald cypress *(Taxodium distichum)* have been destroyed to make mulch. This mulch is reputed to be rot- and termite-resistant, but environmentalists claim that the older trees with those properties are long gone and younger trees do not have the same properties. These coastal forests provide important ecosystem services such as buffering against powerful storms like Hurricane Katrina, which devastated New Orleans in 2005. It is estimated that twenty thousand acres are logged each year in Louisiana alone, mostly for mulch. Make sure you know where your mulch comes from.

Another concern with wood chips and bark is that they can deplete soil nutrients, especially nitrogen. Woody mulches do not have enough nitrogen to support the microbes that decompose their carbon, so the microbes must take some nitrogen from the soil. But while there might be a slight zone of nitrogen deficiency at the soil/mulch interface, studies such as those by Greenly and Rakow in 1995 and Pickering and Shepherd in 2000 have shown that this does not cause much harm. Shallow-rooted annuals or herbaceous perennials might feel the greatest impact, as will gardens where the woody material is incorporated into the soil as an amendment. If you are concerned about a nitrogen deficiency in your garden, put down a thin layer of compost before mulching with wood products and avoid very fresh chips.

In nut-growing regions, crushed shells—from pecans, filberts, and macadamias, for example—are sold as mulch. To reduce carbon emissions in transport, only use nutshells if you live in an area where nuts are grown and/or processed. The shells are very attractive and long lasting but also expensive compared to other mulches. Most organic mulches are highly effective at a depth of at least three inches, so if you don't want to pay for that many shells, you might use fewer as the decorative topping on a less expensive mulch. To appeal to another sense, use cocoa hulls, left over from chocolate production. A deep, rich brown, they smell like chocolate for at least a few weeks after application—which may or may not be a good thing! And if you own a dog, be careful: there is some concern that they are attracted to the hulls, which contain theobromine, a chemical toxic to dogs.

So-called fertile mulches are usually a mixture of bark or wood chips,

TABLE 2. COMPARISON OF COMMON MULCH TYPES

Mulch	Description	Retains Moisture?	Suppresses Weeds?
Bark	Tree bark removed during timber milling	Yes	Yes
Wood chips	By-products of timber milling or wood recycling	Yes	Yes
Straw, grass, pine needles, leaves	Aboveground plant parts	Yes	Yes
"Fertile" mulches	Usually composted manure or biosolids with sawdust or fine bark	Yes	Minimally
Living or green mulches	Densely growing vines or cover crops	No	Yes
Nut hulls	Filbert, pecan, and other shells	Yes	Yes
Gravel, stone	Stones of various sizes	Yes	Yes
Glass	Recycled tumbled glass	Yes	Yes
Cardboard, newspaper	Paper products applied in layers	Yes, but prevents moisture penetration	Yes

which break down slowly, and nutritious but short-lasting materials such as compost or manure. (You can mulch with pure compost, either your own or purchased, but it will break down quickly. See the chapter on recycling for more information.) They are sometimes sold under deceptively delicious names, such as "chicken and chips," which consists of wood chips and manure from chickens and other animals. Some gardeners swear by them, but it is difficult to generalize about them because different manufacturers use different formulas. If you are curious, ask other area gardeners if they use them and which local suppliers they recommend. Ask the suppliers

Reduces Compaction?	Reduces Erosion?	Modifies Temperature?	Sustainably Produced?
Yes	Yes	Yes	Yes, if the trees are harvested sustainably
Yes	Yes	Yes	Yes, if the trees are harvested sustainably
Yes	Yes	Yes	Yes; this waste can be collected as part of regular garden maintenance
Yes	Yes	Yes	Yes
Yes	Yes	Yes	Yes
Yes	Yes	Yes	Yes; best if by-product of local agriculture
Yes	Yes	Yes, and coarser textures are best	Yes, if quarried responsibly; needs fewer reapplications than organic mulches
Yes	Yes	Yes	Yes; needs fewer reapplications than organic mulches
No	Yes	Yes	Depends on amount of recycled material used in production, but using them is a form of recycling, and they biodegrade

what is in their mulch—it will generally be components already discussed, enabling you to judge their sustainability.

Some people use a thick layer of local vegetable matter as mulch. Most gardeners can find pine needles, grass clippings, leaves, and small stems in their own yards. These can work very well and provide some nutrients to the soil, but they will likely need to be reapplied at least annually. If you want to use leaves or stems in particular, investing a bit of money in an electric shredder/chipper will pay off in a short time. Grass clippings are best left on the lawn to replenish the soil with nitrogen naturally. If you live in an agricultural area, you may have access to baled straw—made of barley, oats, or wheat stems—following the harvest. Such fields used to be burned, but this is now banned in most places to preserve air quality, so using the straw helps with a waste problem. Just be aware that straw may contain many weed species.

There is some worry that mulches—and organic ones in particular—may harbor rats and other pests. In general, particulate mulch will not provide a good habitat, but some living mulches, such as dense vines, do offer nesting sites. Sheet mulches, made from materials such as cardboard and plastic, can also provide habitat. But the fears are not often justified: some mulches made from plants in the juniper family, Cupressaceae, contain thujone and other chemicals that repel moths and other insects. (These same chemicals may damage fresh plantings with delicate roots, but they should not harm established plants.) A study by Catherine Long and associates revealed that many bark and wood mulches have fewer termite colonies than gravel. In general, lower-nutrient organic mulches are more resistant to termites, so use them if you live in an area with ground-nesting termites. Dry mulches such as wood chips and gravel may also discourage ticks and other insects.

It is possible that pests such as wood-boring insects will be transported if they are in a tree prior to chipping. U.S. Department of Agriculture scientists concerned about the Asian long-horned beetle infestations in the United States made model insects and ran them through a standard chipper along with wood. None came out intact. However, in the real world, some might leave their tree after transport but before chipping.

More possible is the spread of fungal or viral diseases through processed tree products. Many commercial mulches are heat-treated to kill such pathogens, but arborist chips and other fresh mulches are usually not. It is also possible that chips made from pallets will have pathogens. Disease spread, however, is quite low. Very fresh mulch applied to a susceptible species could transfer the disease (another reason to keep mulch away from the stem), but Linda Chalker-Scott's 2007 survey of peer-reviewed literature on mulch found only a couple of studies that conclude disease is transmitted by mulch and many that conclude it is not. The spatial separation of roots and mulch may help: if the soil is healthy, with beneficial microbes and good tilth, it is unlikely the pathogen will survive to reach the roots. Its survival is more likely if the chips are dug into the soil, which is another reason not to amend soil. Because pathogens vary by region, consult your local extension agent if you are concerned.

Inorganic Mulches

Inorganic materials are used less commonly than organic, but there are many choices available. Although they are usually more expensive, they are also more durable and may cost less over time than organic mulches that need frequent reapplication.

Stones and gravel are very popular, and not just for practical reasons. Some polished stones can give an elegant look to the garden. Gravel is usually smaller in size and inexpensive. But some gardeners hate using rocks because they can be difficult to work with and do not feed the soil as organic mulches do. It is probably most useful for elevating soil temperature. Gravel and stones are generally quarried, so be sure their extraction is properly regulated. Crushed rock, such as 5/8″ minus (a mix with particle sizes from 5/8 of an inch in diameter to very fine), is not suitable as mulch because it packs down and forms a dense and impenetrable layer, but it is better than gravel for walking and driving surfaces.

Recently, tumbled recycled glass has become popular as mulch. It is available in an assortment of colors and can add quite a statement. However, it is probably best used as an accent because it is expensive and so bold that it can easily overpower the rest of the garden. You can create a more subtle look by scattering some among gravel. As you think about where to place it, consider that maintenance can be difficult if lots of leaves fall on it, and it can store a lot of heat.

Black plastic sheeting is sometimes used to conserve moisture and keep down weeds, though due to its unattractiveness it is rarely used in gardens. If the soil is wet and not well drained, plastic sheeting can create anaerobic conditions in the soil that may cause lasting problems for soil chemistry and structure. Thus it is generally not recommended for home use. It is sometimes employed in small agricultural settings to reduce weeds through solarization. Water-permeable weed-blocking cloth, however, is less destructive to soil. Gardeners hoping to remove lawns, a laudable goal, often have better results placing sheets of newspaper or cardboard on the grass and covering them with thick layers of compost. The sheets degrade over a year or so, while blocking sunlight and most water, killing the grass. After the sheets have degraded, the compost is dug into the existing soil and voila! A new garden is born. This method can work well on relatively small, flat areas that are tended and maintained. However, it is subject to some of the same soil chemistry and structure problems as plastic sheeting, and the sheets may also harbor rodents.

Another synthetic material that is becoming more common is recycled rubber tire mulch designed to look startlingly like real bark or wood chips. While it may offer the moisture retention and weed suppression of its organic counterparts, it also delivers things they don't: a huge number of minerals and chemicals used in making and vulcanizing rubber were shown in a 2007 study by Environment and Human Health staff scientists to leach out and persist in aquatic systems, where they have toxic effects. Further, Larry Steward and colleagues found rubber to be the most flammable of the thirteen landscape mulches under comparison. There must be a good way to deal with the vast numbers of tires our car-dependent society produces, but using them as mulch is not the answer.

Final Thoughts

Serious problems arise when we fail to care for the skin of the earth, when we try to fix things that can't be fixed, and when we fail to consider the long-term effects of our gardening practices. Wetter than normal conditions resulted in increased farming in the middle of North America in the 1920s, including the deep plowing-under of native prairie grasses, which for centuries had held soil in place and kept moisture in the soil. Farmers also did not rotate their crops much at this time, and organic matter in the soil became depleted. The 1930s were marked by drought and intense windstorms. By 1934, one hundred million acres had lost most or all of their topsoil, and hundreds of thousands of people left the farms to wander the country in search of work. This eye-opening series of events led to the recognition that soil abuse is a serious matter with long-term consequences.

Gardeners may not face a situation as dire, but the Dust Bowl era is a grave example of what might happen if we do not take care of the land, including our gardens. We should know the soils in which our plants grow and protect them responsibly. Choose plants with your soil type in mind so you do not need to alter it greatly. When you do use amendments, choose them wisely, considering the sustainability of their production. Mulch instead of amending as much as possible, using products that will feed the soil and reduce water use. Ask questions at your garden center and of your extension agents and let them know your concerns. Soil conservation is one of the biggest challenges facing us in the years ahead, and gardeners have an important role to play.

Guidelines

- Don't treat your soil like dirt. Get it tested and check local soil maps before adding anything to it. You likely do not need to add anything, and if you do, you probably need it in smaller amounts than you think. For instance, while organic matter is important for plant growth, it should only be 5–10 percent of the soil.
- Do not dump worms from bins or other sources, especially near natural areas, and do not use commercial mycorrhizae. Practice good soil stewardship to prevent compaction and provide proper nutrient balance, and the appropriate mycorrhizae will find your garden.
- Choose plants that will do well in your native soil. If you have clay soil, it will never be fast-draining sandy loam. Garden centers will help you pick the appropriate plants.
- Do not use amendments when growing woody plants. It is unlikely you will be able to fully amend the soil to the depth of the roots, and even if you could, roots don't grow through mixed soil types.
- Avoid or limit the use of peat and vermiculite. Peat is not collected sustainably at this time. Vermiculite can contain asbestos.
- Pile organic mulches (including shredded or whole fallen leaves) onto beds to naturally feed the soil. Use commercial fertilizers only for containers, vegetable gardens, and other places where you cannot restore fertility naturally through the decomposition of plant materials.
- Use mulch to preserve soil moisture, prevent weeds, and control temperature. Recommended sustainable mulches include bark nuggets, some wood chips, leaves (including pine needles), straw, and grass clippings.
- If mulching to prevent weeds, remove all weeds first. Weeding through the mulch may be harder than pulling weeds from the soil.
- Mulching will not alleviate compaction but can prevent it. Survey your garden for areas where you walk, where you use equipment, or where kids play. Remember that tree roots may extend several feet from the

trunk, so you may need to lay the mulch approximately as far out from the trunk as the branch length.

- Never allow mulch to touch stems. It can create conditions favorable to pathogens, which can easily enter the plant through the stem. Also be aware that arborist mulch may contain parts of trees cut down because they were diseased, though this is unlikely to be a problem unless the mulch touches the stem or trunk of the plant. To keep the soil healthy, do not till wood chips into the soil as an amendment.
- Stones and gravel can be useful mulches to increase water retention and soil heat. Recycled glass is a sustainable inorganic mulch and can add a colorful punch to design, but it is expensive and can be visually overpowering.

CHAPTER TWO

Water, Our Most Precious Resource

We forget that the water cycle and the life cycle are one.

—Jacques-Yves Cousteau

There is one thing of which I am certain: water flows downhill. It is no more able to resist the pull of gravity than anything else on Earth. Unless it evaporates, it will usually find its way into surface water—streams, rivers, ponds, lakes, and oceans—carrying with it much of what it picked up along the way. Since all life depends on water, the effects may be felt throughout an entire ecosystem.

In the past several years, two events have made me much more aware of how my own actions affect the animals dependent on freshwater systems. The first was the listing of native salmon populations in the Puget Sound area of Washington State, where I live, as threatened. By 1999 the survival of Chinook salmon in the region appeared uncertain, and the federal government invoked the Endangered Species Act of 1973, which has instructions on how to treat the habitat critical to the listed species. Chinook salmon spend most of their lives in salt water but enter freshwater to breed. The listing meant that suddenly the many freshwater systems that Chinook could enter from Puget Sound were critical habitats—many in cities, which scrambled to comply with the law. For the first time in the history of this act the burden of protecting an endangered species fell on the shoulders of urban rather than rural residents. All of our actions, all of our sewer and other infrastructure utilities, needed to be examined. Our practices were not the only things hurting the salmon, but they were certainly part of the problem.

I was teaching a class on the ecology of the urban environment at the University of Washington at this time, and I arranged for my students to hear from professionals working to make the Puget Sound region more salmon-friendly and to visit places where the new practices were being implemented. Having a front-row seat to the changes was exciting—it was an opportunity to rethink how we live on this Earth, down to some of the tiniest details that are usually taken for granted. The salmon are still struggling, but the many solutions that were implemented seem to be helping their recovery.

Then, in 2001, I had an even more personal experience after I moved into a house with a garden above a small stream. The stream is not fish-bearing—it originates from springs just a short distance away and empties into a mile-long culvert. Fish cannot get through the culvert, so *my* stream is not considered salmon habitat. But even if fish are not using it, other creatures are. I have Western redbacked salamanders and probably other amphibians as well, which gives me a thrill. It also sobers me to think of my responsibilities to protect their habitat—if something happens to our short stream, my critters will be out of luck. The salamanders are not members of a rare species, but they have a small home range and would not be able to find a new place to live. There could be wider consequences as well: the culvert empties into Puget Sound, which I can see from my house, a daily reminder of the connection between my property—and what I do on it—and the wider world.

A Moment for Watersheds

We all live in watersheds. These are, quite simply, land that surrounds and empties its water into a river, stream, lake, or wetland. One place might be part of multiple watersheds—you could live in the watershed of a stream and also of a nearby river. It is important that each of us know what watersheds we live in and consider them in our garden choices, since they are home to many species of plants and animals and often provide us with drinking water.

When most people think of major surface water systems, they think of rivers, streams, and lakes, but this leaves out a vital element. Wetlands may be less sexy, but they are incredibly important for many reasons. Sometimes called swamps or bogs, wetlands often occur between terrestrial and aquatic systems and can be considered the interface between them—a sort of transition—although they also appear where there aren't nearby aquatic systems. They are usually fairly shallow, and their soil is inundated by water for

The Ramsar Convention on Wetlands

For much of human history, wetlands were considered waste ground and filled in with soil or, often, became garbage dumps. As scientists began to realize the function and global significance of wetlands they became concerned. In 1971 an international group gathered in Ramsar, Iran, to discuss the condition of wetlands worldwide. The Ramsar Convention was developed early that year, and the list of signing nations is now at 159, with nearly 40 billion acres now protected as Wetlands of International Importance. Despite this, the 2005 Millennium Ecosystem Assessment Report speculates that about half of all freshwater wetlands have been lost since 1900, primarily through clearing or draining for agricultural development, but also through diversion, such as for dams.

much or all of the year. They also generally have lots of vegetation. There are numerous types of wetlands, characterized by the species of plants that are typical of them, but all play a significant role in regulating storm water movement and purifying water. They are also very important wildlife habitats.

The Water Cycle

In a way, all water is connected, in what scientists call the hydrologic cycle. Essentially, water does not leave Earth's atmosphere—the water we have today has existed for hundreds of millions of years. It just keeps recycling, as shown in figure 3. Clouds, made of water vapor, send precipitation to earth as rain, snow, or hail, which either evaporates back into the sky or moves in or on the earth. Whether or not water moves into the soil or across another surface makes a big difference. If it falls on soil or surfaces that allow it to reach soil, it either moves straight down to reach aquifers (underground layers of porous earth or stone that hold water) or travels downhill under the soil surface to open bodies of water like streams and lakes. Think of this as water hitting a sponge sitting on a plate—it will be absorbed into the sponge, filling the little pores until there is so much it will pool up on top of the plate.

In the process, the soil "recharges" the water, stripping out the pollutants that it picks up—from the air as it falls, from the lawns covered with fertilizers and pesticides that it drains through, from the roads with oil and other contaminants that it moves across. When it reaches the aquifer or other body of water, it is clean again. The amount of pollutants in rainfall is usually low, and the soil has ways of dealing with them that water does not: pollutants may degrade through chemical interactions on the soil surface or diffuse throughout the earth as water passes through it.

If, however, water falls on impervious surfaces like sealed roads, roofs, and parking lots, it will flow across them, picking up more particles before it

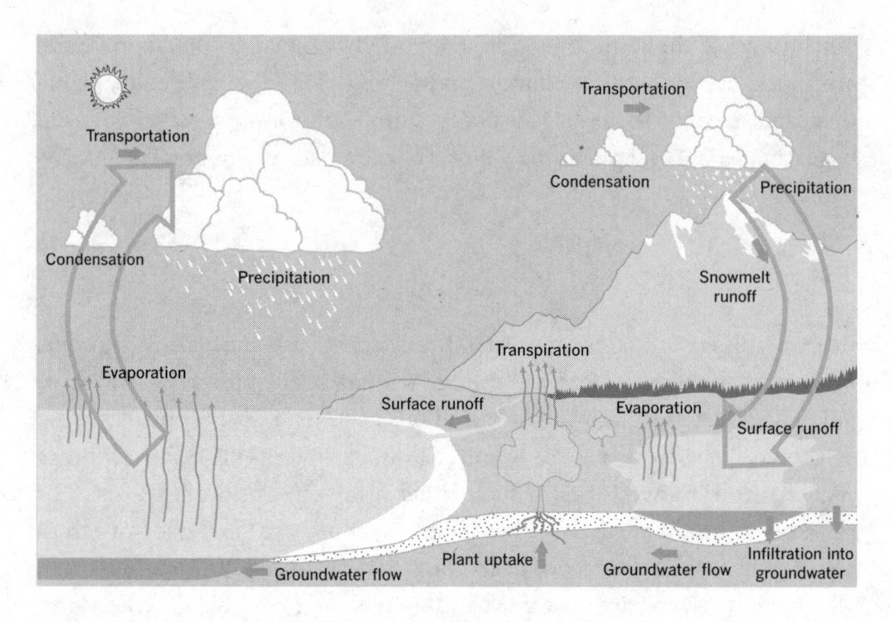

Figure 3. The hydrologic cycle. In areas with extensive hardscape, such as cities, surface runoff is increased and infiltration into groundwater is decreased. Adapted from original by U.S. National Weather Service.

reaches a body of water. This is one way that our rivers, lakes, streams, and wetlands become polluted, and a key feature of urban areas is an increase in these impervious surfaces. If there is a 10–20 percent increase in impervious surface area, runoff doubles; an increase of 35–50 percent, which is typical in some urban areas, can triple the runoff. Those levels usually mean that water reaches open bodies more quickly, before it can shed the bad stuff. Numerous studies documented in the excellent 2001 literature review by Michael Paul and Judy Meyer have linked an increase in impervious surfaces to increased pollution, harmful microbes, and increased water temperature and decreased diversity of nearly all kinds of native plants and animals—and some studies have found that even a 10 percent impervious surface area causes a rapid decline in biological diversity in aquatic systems. Gardeners can increase pervious surfaces by using pervious paving or green roofs or replacing hard surfaces with gardens, as described below.

Pollution from Our Gardens

The chemicals we use in our gardens affect streams, rivers, lakes, and wetlands, whether our gardens are along any of these or many miles away. Pollutants are washed into these bodies of water by both overwatering and rain. Even small doses of pollutants can cause major problems: a 2008 study by Keith Tierney and his colleagues found that pesticides available for home use, such as Malathion, caused problems for salmon in British Columbia in small streams even when the levels were not enough to kill them but enough to affect their sense of smell. This sense tells them when other fish are being devoured by predators, and without this signal they do not know to flee. In 2007, Robert Gilliom found that 90 percent of the water from streams near agricultural, urban, and mixed-use areas had two or more pesticides. In a 2009 study, Cathy Laetz and her colleagues made a discovery that highlights how horrifying those mixtures are. They studied the effect on juvenile coho salmon brains of five common insecticides, alone and in ten pairings and found that the effects were synergistic, with the combinations having 20–90 percent stronger effects than would be predicted by adding the effects of the single chemicals. No fish died from a single-pesticide treatment, but three combinations killed the fish within twenty-four hours. This is bad news indeed for fish—and animals that eat them!

Pesticides

The pesticides we commonly use in our gardens are increasingly detected in urban waters, sometimes at higher rates than in intensive agricultural areas in the United States and elsewhere. Insecticides, herbicides, and fungicides are all showing up. About one-third of all pesticide use in the United States is in urban areas, and Paul Robbins and colleagues suggest that somewhere between 70 and 97 percent of urban households use pesticides. And although agricultural users are required by law to document their application of pesticides, there is no similar law for urban households. Experts call pollution from diffuse sources—such as households in a city—non-point source pollution. Because there are so many contributors, it can be difficult to identify every one and fix the problem. This is especially serious because homeowners often have their pesticides for many years and may be unaware that some have been found to be harmful and may even be illegal. Chemicals like DDT, banned in the United States since 1972, and organochlorine pesticides like chlordane, banned since 1988, are still showing up in urban streams in the 2000s. Even the safer pyrethroid chemicals, which are derived from daisylike plants and largely replaced the banned organochlorines, show up in urban creeks at sometimes toxic levels. The chapter on pesticides gives more information about the effects of these chemicals on animals and natural systems.

Eutrophication

Fertilizers are among the worst polluting culprits—and it does not matter if they are from organic or synthetic sources. Most commercial fertilizers have three key nutrients: nitrogen, phosphorus, and potassium. While these nutrients and others are good for garden plants, nitrogen and phosphorus are not good for freshwater, especially lakes and ponds, as shown in figure 4. When large amounts of these elements wash into the water, algae and aquatic plants increase dramatically. After the initial infusion, there aren't enough nutrients to support all this new life, so many of the plants and algae die. Their decay then consumes the oxygen dissolved in the water, a

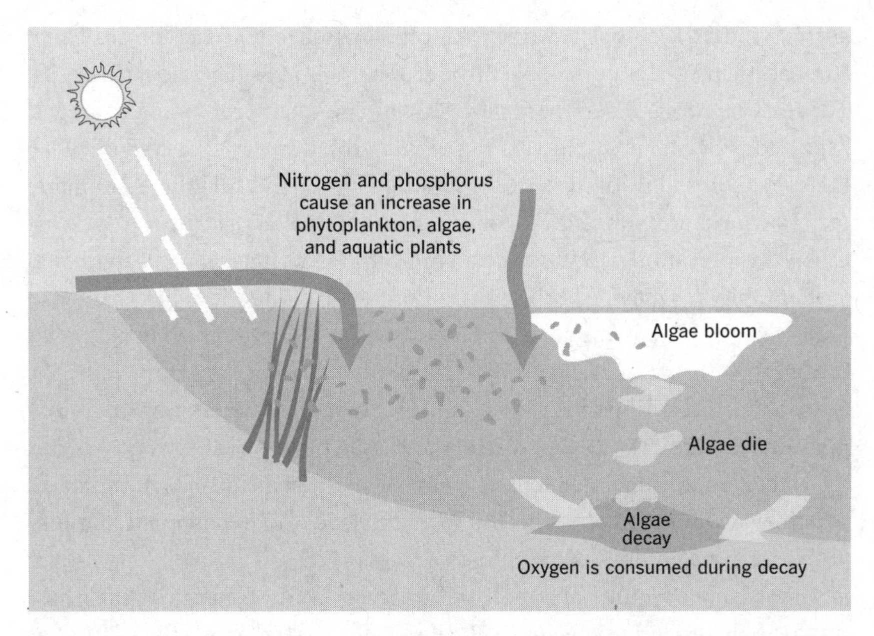

Figure 4. The process of eutrophication. Nitrogen and phosphorus increase the growth of phytoplankton, algae, and aquatic plants. Their eventual decay reduces the amount of oxygen available to aquatic animals, making eutrophic lakes poor habitat. Adapted from original, courtesy of the City of Lincoln, Nebraska.

condition known as eutrophication. If there is less oxygen, fewer fish and other animals can live there. The best conditions for these animals are oligotrophic, in which there are some plants for food and habitat but not a large amount of dead or dying material. Pet waste washing into streams also contributes to eutrophication, so pick up Fido's and Fluffy's "presents" at all times, but especially before storms.

One of the most serious effects of eutrophication has been its worldwide contribution to the rapid decline of amphibians. Amphibians—frogs, salamanders, newts, and their relatives—existed before the dinosaurs and survived after they went extinct. Amphibians have no feathers, scales, or fur, and their skin is permeable. In addition, their eggs have no hard shells. These distinctions make them very susceptible to environmental changes. Many consider them analogous to the proverbial canary in a coal mine, so their decline is alerting us that something is wrong. While there are many factors contributing to this outcome, including habitat destruction and disease, eutrophication appears to be an important one. It has long been known that amphibians may be born with malformed arms and legs in response to parasitism by organisms such as flatworms. Flatworms rely on host species to complete their life cycles, and one important host of flatworms is snails. Eutrophic lakes and ponds often support larger snail populations than oligotrophic bodies of water because of their higher levels of snail food—algae. Therefore, more fertilizer means more algae, which means more snails, which leads to more flatworms, more malformed and maladapted amphibians, and a greater loss of amphibian diversity.

Urban and agricultural runoff accounts for a large amount of the phosphorus in lakes and ponds. In the United States, a study by Jack Gackstatter and associates found that even if sources of phosphorus such as wastewater treatment plants greatly reduced the amount they released into lakes, only 28 percent of the 255 lakes in the study would show measurable improvement. The other 72 percent would not improve measurably unless landowners also controlled runoff from their properties.

Winning the Downhill Battle

If water inevitably drains downhill, how do we stop fertilizers and pesticides from moving with it? The easiest way is to not use them. This needn't come at the cost of your plants. Most woody plants and herbaceous perennials do not require much fertilizer. Mulching with well-aged manure, compost, or other easily broken down organic materials will supply all the necessary nutrients.

If you think a plant has nutritional problems, take a leaf or, better, a small branch with several leaves displaying the symptoms to your garden center, an extension agent, or a Master Gardener for diagnosis. It could be that the soil pH is not right and the plant is not taking up the nutrients that are already available. (See the chapter on soil for more information about pH.) An expert might also suggest adding a little extra phosphorus during the flowering season for herbaceous plants grown for their flowers or for container plantings, which benefit most from an annual refreshing of their soil but might also need small amounts of fertilizer throughout the season.

There are similar simple solutions to make the use of pesticides unnecessary, like putting out a container of beer to lure those notorious lushes, slugs and snails. Dolomitic lime around susceptible plants also deters these pests by scraping up their soft, mucilaginous bodies. Some flying insects are attracted to yellow, so put out yellow sheets of paper or plastic with sticky material to trap them. The chapter on pest control gives many more ideas for avoiding pesticides.

If you feel you must use fertilizers and pesticides, there are a few things you can do to lessen the amount streaming into the, well, streams. Do not use fertilizer or pesticides right before a big storm, and don't overwater after applying them—you will be wasting money in addition to polluting. Some spray-on herbicides have chemicals called surfactants that help them stick to leaves, but the surfactants are often more harmful than pesticides to other organisms! Check labels carefully for surfactants, which are usually banned near water, and to make sure you are getting the correct formulation for your use (the label should say what that is). If you use a granular fertilizer or pesticide on your lawn, keep it off the pavement. If some does land there, sweep it up so water doesn't have a chance to carry it off. And never use water to clean pavement—you can waste more than a hundred gallons in less than thirty minutes, and the gunk it collects can end up in lakes, rivers, and streams.

Designing to Prevent Runoff

Even if you're doing everything right—watering wisely and eliminating pesticides and fertilizers—you may need a way to manage precipitation, which can contribute to runoff and cause erosion. If you get a lot of rain, live on a large property, and have soil that drains well, you might consider creating bioswales to manage runoff. These are long, fancy ditches that you plant with vegetation that can tolerate high water levels. They serve to retain some water, which will then percolate into the ground instead of rushing into nearby streams or down the road. Bioswales can be very attractive and mostly maintain themselves, although they may need to be periodically dredged. They should be relatively shallow, about six to twenty-four inches deep, with a wider bottom (greater surface area) for maximum water filtration and relatively steep sides (up to 4 percent downslope gradient) to hold the water. If the swale is on a slope you may need small dams or weirs to slow the water before it reaches the swale.

You might also consider designing a rain garden into your landscape. These planted areas are similar to bioswales in function but are generally designed as an integral part of your garden. A rain garden should be at least ten feet from the foundation of the house and never on a septic tank drain field. In order to catch and properly infiltrate the water from your roof, its size should be about 20–30 percent of the roof's true surface area, which for some urban gardens is much of the available space. If you have slow-draining soil, such as clay, you may need a rain garden of about 50 percent of the roof area. You can even direct your gutters to discharge their water into a channel (only about a 1 percent slope is needed), lined with attractive rocks, that leads to the rain garden. The rain garden should be at least five inches deep but does not need to be much deeper. Use drought-tolerant plants that can withstand water-saturated soils for a couple of days at a time (check with your local garden center), with maybe a few rocks thrown in for accent, and you will have a beautiful and functional garden feature.

Want another excuse to convince your nongardening partner to let you buy more plants to shoehorn into your garden? Different layers of vegetation—trees, shrubs, herbaceous plants—will slow rain's fall, as drops drip from level to level instead of all hitting the ground surface at full speed and rushing off. This slowing down means less will go sheeting off and more will percolate into the soil and perhaps reach the groundwater. So more plants *are* better! Just allow enough space between them so that air can circulate—you don't want a too-crowded garden to lead to problems like mildew.

What can you do to reduce the impervious surfaces around your garden? New types of pervious concrete and asphalt are being developed, and some are beautiful as well as functional, so as you add or replace garden paths or driveways, explore the possibilities. Flagstone with some room between the pieces is one of the most beautiful footpath surfaces—and the spaces are great places to tuck in fragrant ground covers like creeping thyme, which releases its aroma at each step. You could even break up concrete walkways and re-lay the pieces as paving stones with space between. For an economical option, you might try gravel or bark surfaces, which let water through and discourage weeds.

If you are building a garage or garden shed, consider adding a green roof. Increasingly popular, these use special liners to grow shallow-rooted plants (usually drought-resistant species). The earliest roof gardens may have been the Hanging Gardens of Babylon, one of the seven wonders of the ancient world. This wonder has gone high-tech in the modern world, and there are growing numbers of landscape designers specializing in them. These roofs not only absorb or slow down much of the water that would otherwise rush off and down the driveway but also store water during storms, which delays runoff until after the rain has peaked. Any water that does not run off returns to the atmosphere through evaporation. Depending on the design and other factors, a green roof can reduce annual runoff by 60–79 percent. The blanket of earth and plants also increases the building's heating and cooling efficiency. Entire houses and even factories have green roofs, so once you have done your garden shed you may be inspired to put one on your home.

But maybe you would rather collect the water from your roof to irrigate your garden, since a roof of fifteen hundred square feet can provide twenty-two thousand gallons of free water each year. Many garden centers now sell water barrels that hold about fifty-five gallons and can be hooked to the pipes that drain our roofs. You can even attach a hose to the spigot and use gravity to water the garden. Some barrels are designed to be hooked together so that when one fills, the water overflows to the next one, and so on. Because different parts of the roof drain to different downspouts, you might want several. All should have sturdy covers to prevent children and small animals from entering. Some are ugly, but if you shop around you should find some that suit your house.

People are also employing cisterns for water collection. Cisterns are larger tanks capable of storing thousands of gallons of water collected from the same surfaces as barrels. They originated for household use in the Middle East before 1000 BCE and perhaps as early as 2000 BCE. In some areas, it was decreed that all households *must* have a cistern for private use. It appears that people back then managed their water better than we do today. But modern cisterns are improving on the age-old technology, and some landscape designers and installers now specialize in them. Cisterns are still usually simple above-ground structures, but increasingly high-tech ones that hold water funneled from impervious surfaces are being buried beneath the ground. (Many people take advantage of the removal of a buried heating oil tank to install them, but a hole can be dug especially for this.) While cisterns can be expensive, not all are, and they provide an excellent way to store water for later landscape use.

A River Runs through It

Those lucky enough to live along lakes, ponds, streams, and rivers may enjoy the wildlife they attract, but these people also have special responsibilities because their close proximity means they could do the greatest harm though pesticide and fertilizer use, among other things. Lakes and ponds are among the most vulnerable of our freshwater bodies because they usually do not cleanse themselves with new inputs of fast-moving water. They act as repositories—or sinks, as scientists call them—of a wide variety of contaminants. They are isolated ecosystems similar to islands, which are particularly susceptible to invasion by nonnative plants and animals. They each have their own balance of native plants and animals, and the introduction of a nonnative species, even one that is native to a nearby place, can disturb the balance. You should *never* dump an aquarium into a body of water or undertake to stock a lake or pond with a favorite animal or plant. (See the chapter on invasive plants for more tips.)

Streams and rivers have special needs too. Before making any major changes to your stream or river you should check to see if you need a permit. Never dredge or try to alter the channel of a stream without contacting authorities first—there can be serious legal and ecological consequences. Also refrain from removing woody branches and logs, which serve many functions. They slow and direct the water, supply important nutrients, and

Go Play Outside . . . Carefully

I remember hunting in streams for crawdads, a sort of mini-lobster, and other cool critters while growing up in North Carolina. I have no doubt that those investigations fueled my desire to become a biologist. But too often today, children stay inside with video games and never develop curiosity about the natural world. They should be encouraged to explore and imagine outside. But we must also teach them about climbing in and out of natural water systems so that erosion is minimized and about respecting the plants and animals that depend on the water for survival. Stomping around in water is fun, but it's best done in constructed water areas that aren't home to important organisms.

provide hiding places for fish. Instead of thinking of them as untidy, think of them as habitat for many species of desirable wildlife!

A general concern for those who live next to water is soil erosion. Too much soil entering the water can increase the turbidity of the water and decrease water quality. Soil will also fill in the porous areas among the rocks along the bottom, which are critical to many species and some spawning fish. There is a temptation to reduce erosion by placing rocks or other hard objects along the stream bank, but this can make the problem worse. The hard surfaces can deflect water currents to hit the banks nearby, increasing erosion.

Plantings will help keep the soil in place and limit access to the water. They can also provide habitat and food for wildlife. Native species are generally the best to select for this, but a mix of natives and noninvasive nonnatives might be acceptable. Remember that leaves, flowers, and other parts of the plant will fall into the water, so choose species that are not toxic and that will not dump lots of plant parts at once, contributing to a lack of dissolved oxygen as they decay. It is best not to have lawn directly above water because the fertilizers and pesticides essential to create lush green grass are highly likely to wash into it.

Xeriscaping

There is not enough clean freshwater for Earth's growing human population, and the United States is not safe from this problem. Sometimes geography exacerbates the situation. Southern California is perhaps the best example of this—Los Angeles normally gets only three inches of rain a year, but the desirable climate means that people flock to live there, creating more demand for water than can be locally supplied. So the city gets much of its water from Arizona, through a 242-mile aqueduct. I live in an area known as the rain capital of the country, but some years we worry about having enough water for drinking and other basic functions. To help with this growing concern, gardeners in all parts of the country, even the wet ones, should consider their water conservation practices.

Xeriscaping, or xerophytic gardening, is a type of gardening that aims to use as little water as possible by combining careful plant selection with water-saving irrigation techniques. It gets its name from the Greek *xero,* meaning "dry," and the word *landscaping.* The concepts are fairly simple and straightforward.

Plant selection is the place to start. Use mostly plants from your region that do not require much irrigation, putting high-water-use plants only in garden areas that are naturally wet. Group plants with similar water needs to make irrigation more efficient. Local experts such as garden center employees and Master Gardeners can help locate plants suitable for your region that are hardy and do not use too much water. You'll need to consider not only your annual amount of rainfall but also when it occurs. Some places have rain throughout the year, while others have more in the winter or the summer. People in areas with more precipitation in the winter than in the summer must be especially careful in their selection because even many drought-tolerant plants need extra water in the growing months.

Reduce the amount of water-hogging lawn, or allow your lawn to go brown when it is dry (most grasses can survive extended droughts, even if the aboveground parts turn brown). Or try one of the ecoturf selections that require less intensive maintenance. These lawns resemble a meadow and have a mixture of grasses, waterwise clovers, and other perennials. They take a year or two to establish but then usually need much less water, fertilizers, and pesticides to maintain a healthy appearance and require relatively infrequent mowing with a mulching mower. The best combinations of species to use will vary widely by location, so check with local extension services, Master Gardeners, and garden centers for advice.

Then consider your watering practices. It is generally best for plants if you water deeply and less often: usually about an inch of water, including precipitation, each week is enough. (To monitor your rainfall, set an empty, straight-sided, shallow container, like a tuna can, outside and check to see how much water it gets.) Use efficient irrigation systems that sense precipitation if they are on timers, so you do not water unnecessarily. Soaker hoses, drip systems, irrigation systems with zones for different watering needs, and systems that sense rainfall and do not water during or right after a rain (how often have we all seen that?) are all good ways to reduce water use, sometimes by as much as 30 percent. You want to avoid seeing your irrigation water flowing down the street, a condition sometimes called urban drool. That is wasted water (and money) and a good way to pollute local waters.

Do what you can to prevent evaporation of water from the soil surface. It's a good idea to water in the morning or evening because less will evaporate during these cooler times of the day. Ground cover can also help. Mulch not only prevents the evaporative loss of water from the surface but also slows the growth of weeds that will steal the water in the soil from the plants you want. A downside of mulch is that it takes more water to penetrate to the soil surface—but the positives outweigh this small negative. The chapter on soil contains more information about mulches.

Final Thoughts

As I write this, I am gazing out across Puget Sound, a complex of marine waterways into which run countless streams and rivers. It is a beautiful clear summer day and there are sailboats and people fishing. It looks idyllic. Four million people live in the watersheds that drain into it, however, and what is happening beneath the surface of the water is far from idyllic. Gilbert Bortleson from the U.S. Geological Survey and Dale Davis from the Washington Department of Ecology found that creeks in the central Puget Sound area had two herbicides and one insecticide that are commonly used by homeowners at every test site, probably resulting from stormwater runoff. Experts consider runoff to be the single biggest threat to the animals and plants that depend on Puget Sound. Near-shore plants are dying, and the juvenile salmon that use them to hide from predators are dying too. Shellfish beds are dying, orca whales are declining, and eutrophic "dead zones" are increasing.

This story is, sadly, not unique. The Chesapeake Bay in the Mid-Atlantic, the Great Lakes in the Midwest, the Everglades in Florida—all share this tragic tale. What water body does your watershed drain into, and what shape is it in?

Experts know that, as with almost every conservation issue, science is only part of the solution. The bigger part is social: how do we get people to understand the consequences of their actions and modify their behavior? The listing of Puget Sound salmon as endangered is opening the eyes of people in this region to the changes they need to make. You can follow the small and large steps in this chapter to make your own changes and protect *your* waterways before they become the next Puget Sound.

Guidelines

- Find out what watershed you live in. Your water utility company or even your local Parks Department might be able to help. Investigate opportunities to help restore and maintain it.
- Do not use more fertilizer or pesticides than are absolutely needed, and never apply them before a storm is predicted. If granular material gets on the pavement, sweep it up. Never use water to clean debris from pavement.
- Add more plants to your garden, staggering them in canopy layers to slow rain as it falls. Design rain gardens or bioswales to hold water until it can seep into the ground. Use a green roof on smaller garden sheds, playhouses, or even part or all of your home's roof to slow stormwater runoff.
- Replace impervious with pervious surfaces as best you can. Using flagstones or the many new permeable pavements for paths and patios is a good start.

- Select plants that use less water and add mulch around them to prevent water loss as well as discourage weeds.
- Add rain barrels or cisterns to collect the water from your roof and other impervious surfaces to use for irrigation.
- Examine whether you really need a lawn, which almost always requires additional irrigation, fertilizers, and pesticides. If you absolutely can't live without one, investigate ecoturf options suited for your climate.
- Learn about Adopt-A-Stream, catchment associations, and other stream stewardship programs in your local community. Your water utility company can probably refer you. Attend some information sessions or even volunteer.
- Limit access to natural water bodies to prevent erosion and to protect aquatic animals. Allow logs and other materials to remain as long as they are not blocking the flow of water. If you think a log should be moved, contact local authorities and find out if you need a permit.
- Never dump aquaria in or otherwise add nonnative plants or animals to natural water bodies. Do not even add them to an artificial pond if there is a chance it might overflow into a natural body of water.

CHAPTER THREE

Should You Go Native?

What better expresses land than the plants that originally grew on it?

—Aldo Leopold, *A Sand County Almanac*

A. Starker Leopold, son of the great conservationist Aldo Leopold, presented a landmark report in 1963 as chair of the Advisory Board on Wildlife Management for the U.S. National Park Service. This work, often referred to as the Leopold Report, drew on the senior Leopold's philosophy to conclude that "the goal of managing the national parks and monuments should be to preserve, or where necessary to recreate, the ecologic scene as viewed by the first European visitors. As part of this scene, native species . . . should be present in maximum variety and reasonable abundance."

The Plant Conservation Alliance, a group of ten federal agencies and hundreds of nonfederal collaborators, calls a species *native* if it "occurs in a particular region, ecosystem, and habitat without direct or indirect human actions. Species native to North America are generally recognized to be those occurring on the continent prior to European settlement." The indigenous peoples of the Americas did move plants around, both intentionally (for example, corn, a staple of the diet in eastern parts of North America when Europeans arrived, descended from maize brought from Mexico) and unintentionally (as they made seasonal and long-term migrations they most likely carried some hitchhiking seeds). But the plant migrations they effected were on much smaller scales of number and distance than those of later eras. Other parts of the world, like New Zealand and Australia, also define *native* as those species that were present before the arrival of Europeans. Even Europe, with its longer history of human habitation and movement, takes a similar approach. There, species are called *archaeophytes* if they were introduced prior to the Columbian Exchange with the New World starting in 1492 and are treated essentially as natives, and *neophytes,* likewise treated as invasives, if they came later.

Geographic scale also enters into and complicates the definition of *native.* While species may be native to an area where they are sold, they may also be carried in stores in regions where they do not naturally occur. For example, sugar maple *(Acer saccharum)* is native to the eastern United States, but it is also grown and sold on the West Coast, where it is not native. In addition, some species are wide-ranging, even spanning continents, such as circumboreal species, which are found at high latitudes around the globe. Shrubby cinquefoil *(Dasiphora fruticosa,* also known as *Potentilla fruticosa)* is native to high latitudes and elevations in North America, Europe, and parts of Asia. European genotypes are grown in North America—should they be considered native plants here? Be careful to accept a claim of "native" when purchasing a plant—ask where it is found in the wild. If it is a named cultivar, Internet searching may show the source of the clones.

The potentially maddening issue of defining *native,* and other words like it, has become increasingly important over the past two decades as gardeners have shown a growing interest in using native species in landscape plantings. The claims about natives at times have become passionate exclamations of superiority, with supporters of natives and nonnatives each arguing with the fervor of a Baptist minister at an old-time revival. Advocates say native plants are more suited to the local environment, interact better with native wildlife, require less pesticide use, and are just plain prettier. They are also integral in imparting a sense of place. Detractors say that while native plants are just fine, they do not want their palette limited to just those species, and that growing native plants in gardens might harm the long-term survival of some native species through gene mixing and unsustainable collection. In general, the use of native plants in the landscape should be encouraged, but some of the assumptions about them need a critical evaluation. In addition, there are conservation issues that gardeners should consider in making their selections.

Are Native Plants Better Suited to the Local Environment?

There is an understandable assumption that native plants will grow better, with less effort, than nonnatives: because they evolved locally, they should be better adapted to local conditions. But our native lands are a mosaic of different ecosystems—wetlands, rivers and streams, forests, grasslands, and so on—and a species that grows in the wetter spots of a dry region may not be as water-efficient, for instance, as many assume. In addition, most of us are urban or suburban—we are not gardening in forests or prairies, so our local ecology is different from that of the surrounding wildlands. In any location, there are native plants that can cope with altered conditions, but a nonnative might be a better choice. When making the decision, be sure to consider the soil, temperature, and availability of water in your area.

Soil

Soil is made of particles of rock, fungi, decaying organic material, and, importantly, space (see the soil chapter for a more detailed description). The porous space in soil is occupied by water and oxygen, two critical resources for plants. When heavy machinery runs over and compacts soil—during building construction, for instance—the pores are reduced, affecting the ability of plants to grow and thrive. Lots of people walking in a restricted area, such as on a path, can also compact soil. Mulch can help with the latter, but heavy machinery is probably too weighty for mulch. Even when the construction process does not badly compact the soil, it may remove the top layers, which contain most of the organic matter that plants need. Compaction and a thin layer of organic matter may be hard for both natives and nonnatives to deal with. There are very few plants, such as David's viburnum *(Viburnum davidii),* that can grow under these conditions, and it is possible that your region may not have suitable natives. Garden centers can help with plant selection for compacted soils.

Another soil factor important to plant growth is pH. It is measured on a scale from 0 to 14, with numbers lower than 7 representing acids and those above, bases. Most plants prefer neutral or slightly acidic soils, but a few species grow better in basic soil. The pH of the soil is raised by lime, a central component of concrete. And what is found in urban areas, from sidewalks to house foundations to planter boxes? That's right—concrete! As concrete decays over time, lime leaches out, and the soil near it becomes more basic. The physiology of each plant species will determine its tolerance of decreasing soil acidity, but natives may deal with pH changes no better than nonnatives.

Temperature and Air Pollution

Urban areas often have hotter climates than surrounding areas (see figure 5), so much so that scientists refer to cities as urban heat islands. This was first noted in the early 1800s by Luke Howard, the British scientist responsible for the basis of the cloud classification system we still use. He observed that the center of London was warmer than the exterior. In 1997 the National

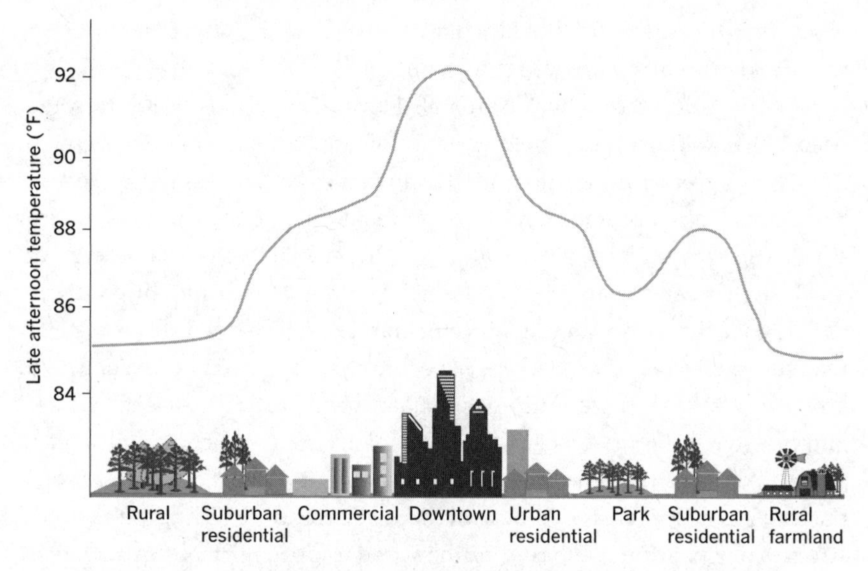

Figure 5. Temperatures are generally higher over landscapes with extensive built environments. Adapted from original by Natural Resources Canada.

Aeronautics and Space Administration (NASA) and the Environmental Protection Agency (EPA) began the Urban Heat Island Pilot Project to monitor temperature and precipitation in major U.S. cities with satellites and ground-based measurements. The EPA found that the seventy cities studied in eastern North America are on average 4.1°F warmer than the surrounding areas in the spring. Although there has been much debate, according to the Intergovernmental Panel on Climate Change (see chapter 7, on climate change) the urban heat island effect is unrelated to global warming—but it is a local problem. These temperature differences should be within the tolerance range of most native species, but they may add to the stress on some: urban plants may start growing leaves about seven days earlier than in surrounding areas and hold them for another week in the fall, requiring water and nutrients longer than normal.

How do cities get so hot? Although Howard's early work led to some useful observations, the development of techniques such as remote sensing created greater understanding. For instance, NASA began a project in Atlanta in the 1990s that found that the vast built surfaces in urban areas, such as streets, sidewalks, and buildings, absorb and hold solar energy at much higher rates than trees and soil—and sometimes measured up to 70°F higher than surrounding air temperatures! This stored heat is slowly released, increasing the air temperature up to 10°F above that in suburban areas. In addition, these buildings usually have heating elements, whose output even in energy-efficient buildings does not completely stay inside, and air-conditioning units that generate heat. The photochemical reactions that cause smog intensify at these higher temperatures, adding yet another stress to plants through blocked sunlight and trapped heat. NASA scientists believe that the rising hot air from cities (the heat island extends upward by three to five times the average building height) creates or enhances existing clouds. Combined with the additional particles in the air from pollution, this means urban areas may get 48–116 percent more summer rain and affect vicinities up to 40 miles downwind by increasing precipitation by about 28 percent per year. Nonnative plants may tolerate the increased temperatures, air pollution, and precipitation better than natives that are used to a different environment.

Water

Some enthusiasts believe that native plants use less water because they evolved in the region and are adapted to the precipitation patterns. Plants do generally grow best in areas with the conditions they evolved under, or close to them, though adaptability varies widely by species. Certainly, growing a species from an area with year-round rainfall in an area where precipitation mostly happens in the winter is likely to mean that irrigation will be necessary in the summer: a species from China, which commonly has summer rainfall, is likely not the most waterwise choice in a Mediterranean-type climate like California's. Would a gardener in that state be more waterwise to grow a species from the Mediterranean, Chile, or another area with a similar rainfall pattern?

Few studies have documented the actual water use and requirements of landscape plants. There are many lists of species that use less water, generally developed through the observation of plant survivorship under arid conditions—but although many plants can survive drought stress, some may be high water users when the resource is less limited. This type of analysis has mostly been used for crop species, including fruit-bearing plants, but there are some cases where it has been used for landscape plants. Daniel Levitt and colleagues, in a study done in the arid state of Arizona, compared water use by a nonnative oak cultivar with that of mesquite, a native desert shrub. When water was plentiful, the mesquite took up considerably more than the nonnative oak did. Rita Hummel and Ray Maleike showed that *Escallonia,* native to Chile, has water use similar to that of a native of the western United States, madrone. Both areas have periods of summer drought and winter rain, and these two very different species have evolved to cope with those conditions. But the Hummel and Maleike study also looked at rhododendrons and found that while some introduced species use more water than native species, some use less. The study concluded that geographic origin is not a good predictor of water use when water is not limited. If you irrigate your garden for species that require more water, assuming that the drought-tolerant native won't be taking much of it, you may find your water bill higher than expected because the native species *is* using it. You might want to place the native in a part of the garden where you will water less.

Are Native Plants Better for Native Animals?

Another argument commonly made for native species is that because they evolved with predators such as native insects, they are less likely than non-native plants to require pesticides. Every plant species has something that eats it, whether specialists (adapted, perhaps chemically, to a particular species), generalists (able to consume a wide variety of plants), or both. The balance of specialist to generalist native herbivores can affect the amount of damage to the plant. There is a field of scientific study called biological control that is based on the idea that specialist insects and diseases native to the same part of the world as an invasive plant are the best choices to reduce the vigor of that plant. Biological control experts will go to the native range of a weed or pest insect and find native predators to introduce in the invaded area to keep it in check, in keeping with the enemy release hypothesis, which posits that species become invasive in places where they are introduced because their coevolved predators are not present. But just because the plants now occur in the same area as their predators does not mean they evolved together—each species migrated into the area on its own, maybe centuries apart from the others and with characteristics perhaps already set in response to different conditions. Even if native plants are adapted to native insects—and there is no question that native plants, like most plants, have evolved mechanisms to avoid herbivory—new generalist pests are often introduced too. For instance, a new pathogen was detected in California in 1995: *Phytophthora ramorum,* the cause of sudden oak death, can affect and even kill many other species. It is certainly true that some plants are less pest-prone than others, but little scientific evidence suggests that native plants in general are among those. In fact, Kristina Schierenbeck and her colleagues found that when a native and a nonnative honeysuckle were exposed to natural predation, the native suffered considerably more damage and the primary insect predators were native.

Whether plant-eating insects and plants have coevolved is still hotly debated. Insects have relatively short generation times, so they can respond rapidly genetically to opportunities for subsisting on different plant species. Plants cannot respond to insects as quickly because of their much longer generation times, but it is believed that herbivorous insects have influenced the evolution of plants and that the plants' adaptations triggered corresponding evolution in the insects. One frequently cited but problematic model is the monarch butterfly and the milkweed plant. While adult monarch butterflies are able to forage nectar from a variety of plants, as caterpillars they only eat milkweed *(Asclepias tuberosa).* The milkweed, which as the common name implies has a milky sap, produces chemicals that are distasteful to birds. The caterpillars ingest these chemicals, which stay with them as they grow into adults and some of which they pass on to their eggs. Birds have learned that despite their succulent appearance, monarchs are not dinner material. However, since the plants do not receive benefits, this is not really coevolution.

The bottom line is that assuming that native plants and animals coevolved and that native plants are better for a landscape is questionable at best. It is doubtful that many of them actually did evolve in response to one another, and little evidence exists to suggest that there is less insect predation of native plants and less need for insecticides. Trying to match coevolved species is not a good reason to use native plants.

However, even without closely linked evolution, native plants are often best for native animals. Birds find nesting sites using visual searches for characteristics that will keep the nest safe from predation. Nonnative plant species may have some of these characteristics, but they might also have negative attributes—which birds might not notice. For instance, Kenneth Schmidt and Christopher Whelan compared nest predation in native and nonnative shrubs in the U.S. Midwest over six years and found that American robins *(Turdus migratorius)* and wood thrushes *(Hylocichla mustelina)* that nested in nonnative honeysuckles and buckthorns lost more young. The researchers attributed this to the nonnatives lacking thorns (found in one of the native species) and having lower branches, forcing the nests to be closer to the ground and therefore vulnerable to a wider range of predators. In addition, in gardens near streams, rivers, and lakes, planting natives near the water may provide superior food sources and habitat for fish, salamanders, and other aquatic or semiaquatic wildlife.

Using native plants may also help prevent the distribution of invasive species. Birds spread about half of all woody plants by eating their fruits and regurgitating or defecating the seeds. If a native species is on the menu, they will be more likely to disperse it than an invasive.

Does Growing Natives in Your Garden Harm Wild Populations?

Most species are genetically variable, and even a single population may have many different genetic forms. This is often seen in slight variations in flower color or plant height. Even if individual plants look the same, one may have genetic differences that make it better adapted to local conditions, such as being able to survive with less annual precipitation. Species with large geographic distributions and those that depend on cross-pollination from other individuals, in particular, may have very diverse genetic forms, or genotypes.

Decline in Genetic Fitness through Outbreeding

Propagation nurseries may use seeds to produce new plants, but often, especially in the case of named cultivars, they propagate cuttings from stock plants. Genotypes acquired from commercial businesses such as garden centers may not be native to an area, even if their species is. The evergreen ground cover kinnikinnick *(Arctostaphylos uva-ursi)* is a good example. It is found across the higher latitudes of North America, from the West to the East Coast, and Eurasia. It is heavily marketed in the Pacific Northwest because of its drought tolerance. There are very nice cultivars, such as

'Vancouver Jade', developed from the region's native plants, but one of the most commonly sold cultivars is 'Massachusetts'. Although it was selected by a horticulture professor in the Pacific Northwest, the source of its seed is the namesake state, which gets substantial summer precipitation, unlike the West Coast. The concern is that the genes of the eastern genotype can get into wild populations in the West. Wind, insects, and birds spread pollen from one plant to another, and even if it starts in a garden it can end far away, perhaps on a wild plant, and establish new, hybrid populations, sometimes near native wild ones.

Outbreeding is the term used when different genotypes in a single species that are usually unable to reproduce because of spatial separation are brought closer together, as can happen through horticulture. When such hybridization of genotypes results in less-fit forms of the species, it is called outbreeding depression. Inbreeding depression might be more familiar to many people: when closely related individuals breed, the offspring might express more recessive traits. A classic example of inbreeding depression is the higher prevalence of insanity, hemophilia, and other genetic problems among the royal families of Europe—because they have historically intermarried for political reasons, the chances of their expressing genetic disorders are higher. But with outbreeding depression the genes introduced by a genotype that is not native to an area are less adapted for the local conditions and can cause serious problems. Other, more descriptive terms sometimes used for this include *gene swamping* and *genetic pollution.* If the wild population is small, the chance that it will vanish because of even a temporary change in suitability to a site is greater than in a large population, which can assimilate the changes.

How likely is it that outbreeding depression would occur? It is hard to say because it depends on many factors. If the number of introduced individuals is high, as in a mass planting or an agricultural setting (including a large wholesale nursery) or if gene flow between wild and domestic populations through efficient pollinators or seed dispersers is high, the chances increase.

It can also be more of a problem, as noted, if the wild population is small, so the nonnative genotypes are more quickly integrated. Norman Ellstrand and his colleagues found that of thirteen of the world's most important food crops, twelve hybridize with wild relatives in some part of their agricultural range—which can have important implications for rare species. Large ecosystem restoration projects can also cause outbreeding depression. Arlee Montalvo and Ellstrand concluded that mixing genetically different seed sources of common deerweed *(Lotus scoparius)* lowered the fitness, or potential of the population to reproduce and survive in the future, of augmented or restored populations. Forest plantations of native trees, a form of agriculture, may also have this effect—Brad Potts and his colleagues in Australia found that plantations of nonlocal eucalyptus put wild trees at risk.

While scientists argue about the importance of outbreeding depression, it is too early to tell if this is a concern in most smaller-scale horticultural settings, so caution is best. In general, you should not grow a species that is found in nearby wild populations (how nearby is dependent on the method of gene movement, but as a rule of thumb a mile is a good estimate) unless you can use plants propagated from those wild populations. If you are creating a mass planting, which might attract more pollinators and increase gene flow, you should be sure you are an even greater distance away from local wild populations. Consult a native plant society to determine what is native in your area.

An interesting trend in horticulture is the introduction of naturally occurring cultivars that are carefully selected and bred for certain characteristics. They may have an unusual flower or leaf color, growth form, or other variation from the wild types native to a region. For instance, a double-flowered form of large-flowered trillium *(Trillium grandiflorum)* rarely occurs naturally but is now grown in gardens. A white-flowered form of pink-flowered currant *(Ribes sanguineum)* is popular but does not attract pollinating hummingbirds, which are needed to help pollinate the plant but are not drawn to white flowers, causing the currant to produce fewer seeds. While these

"California" Poppies

The state flower of California, the California poppy *(Eschscholzia californica)* is an excellent example of how horticulture can influence wild populations. S.A. Cook studied the population biology of the poppies in the 1960s and found that the species has wide habitat-correlated variation, perhaps reflecting genotype adaptation to local conditions. The poppies are native from northern Oregon to central Baja California and from the foggy coastal ranges to the inland dry areas, from sea level to sixty-five hundred feet. The individual plants are not sexually compatible with themselves, so they require insects to carry pollen from one plant to another in order to produce seeds. This also suggests that they are highly genetically variable. Cook showed that California poppies in coastal areas tend to be perennial, putting more effort into producing long, robust tap roots than flowers, fruits, and seeds. The seeds tend to have no dormancy and germinate easily right away. Poppies in drier inland areas tend to be annuals, with weaker roots, more flowers and fruits, and seed dormancy. This life strategy is common in species that live in highly variable environments, where being able to handle erratic weather conditions is an advantage for survival. The coastal plants, with deep, thick roots supporting evergreen leaves, are better adapted to more stable climates. But these differences in life strategy persist even when the two types are grown together in a greenhouse, proving that they are genetic and not environmentally induced. Crosses of the annual and perennial forms have less-fertile pollen, at least in the first generation, resulting in fewer seeds. However, packages of poppy seeds are widely available to home gardeners in both poppy-only packages and wildflower seed mixes and are not produced or labeled with local adaptations as a consideration. And here is an even more surprising fact: although the species is native, the seed sold in California is mostly grown elsewhere, from a limited genetic source. The California Transportation Department told Gordon Fox, who has studied the poppies, that all of the seeds its workers use for roadside plantings come from one source in the Netherlands, and it is likely most garden seed does as well. Fox and other botanists are concerned that the Netherlands form is hybridizing with wild populations in the United States.

botanical curiosities are great fun in the garden, you should consider their possible ability to hybridize with local plants. If wild genotypes are found nearby, it would be safer to choose another plant.

Preserving Rare Native Plants

From a biological standpoint, outbreeding depression is even more important with rare native plant species. They may be rare because there are only a few populations, so the introduction of even a small number of less-adapted genes could mean the loss of one or more populations, which could be very serious for the long-term survival of the species. Or the populations might have only a few individual plants (as I have already noted, small populations are at greater risk of disappearing forever). As with more common native plants, the threat of gene swamping really depends on each gardener knowing what is growing wild near his or her garden.

Beyond the biological concerns, there are political reasons for not growing rare native species. Scientists who work with rare species outside of wild populations, which is known as *ex situ* conservation, always worry that politicians will think that if plants can easily be grown in horticultural settings, there is no point in saving the places where they grow naturally. Politicians may reason that while it is important to save a species, it can be saved in a garden. But these artificial conditions are no substitute for the real thing, where the species are connected with the others that they depend on or that depend on them and where they can be grown in large enough populations to prevent inbreeding depression. It is always best that rare native plant species be left to grow in the wild or grown only as part of carefully designed and coordinated programs.

Gardeners can play a role in protecting rare species, and some species would have been lost if not for their presence in gardens. A classic example is *Franklinia alatamaha,* a camellia relative that is now extinct in the wild but, due to its presence in botanic gardens and, to a lesser extent, home gardens, not completely gone. John Bartram was a Pennsylvania farmer who, according to his son, one day stopped plowing a field to ponder the beauty of a daisy and thereafter became an acclaimed botanist. King George III of England named him the Royal Botanist for North America in 1765, and later that year he first described the *Franklinia* from plants he found growing along a river in Georgia, naming it after his friend, the scientist and philosopher Benjamin Franklin. Bartram's son, William, later collected and sent seeds from the plant back to England. *Franklinia* has not been seen in the wild since 1803, so all examples of it now are descended from the ones Bartram collected. A survey done from 1998 to 2000 by the staff of the Bartram Garden (yes, it still exists!) to mark John Bartram's three-hundredth birthday found more than two thousand of these plants worldwide. Most were in public and private gardens in the United States, but many were reported in eight other countries.

Although Bartram's actions inadvertently saved this species, its story is really pretty sad. Wouldn't it be better if we could still go visit the wild populations that Bartram first saw? And who knows how the loss of this species affected other organisms along that river? The species was saved, but its community is gone.

Saving the seeds of rare plants is important, but collection must be done carefully to preserve the wild populations and to ensure genetic diversity in the collections. Many botanic gardens bank such seeds and allow trained volunteers to help—check with your local gardens if you are interested.

Could plants like these in garden refuges around the world be used to reestablish a species that is critically rare or even extinct in the wild? Possibly, but the new population would be a poor substitute. Wild populations contain a number of genotypes, but nursery plants are usually propagated from a limited number of mother plants. If we put plants from that stock in the wild, we would be planting just a tiny representation of the genetic material of the population and of the species as a whole. As we saw earlier, a lack of genetic diversity leaves plants vulnerable to changing conditions in the wild because the chances of there being a genotype suitable to the change are reduced. Trying to establish a new population of *Franklinia* from the descendants of Bartram's collection would therefore likely fail in the long term. Plants in gardens are only a very small, artificial representation of their species—but if the worst happened, they could mean the species would not be totally lost.

Controversy over Wild-Collected Material

Wild collection of plants of all sorts—both native and introduced—can lead to serious conservation problems. During 1988–1989, more than twenty-six million bulbs of the giant snowdrop *(Galanthus elwesii)* were imported to the Netherlands from Turkey for eventual distribution throughout the world. It was the practice of Turkish villagers to collect wild bulbs, which soon became scarce. Large bulbs were sent for export, and the smaller ones were grown in fields in Turkey until they were big enough to ship. The latter were then labeled as propagated, even though they were originally wild-collected. Today, this would not be legal under the Convention on International Trade in Endangered Species (CITES), a treaty signed by 144 countries that regulates how threatened plants and animals can be used in trade to prevent their overexploitation in the wild. Under its rules, parental stock must be obtained in ways that to do not harm the survival of the species in the wild and managed in such a way that the long-term maintenance of the parental stock is guaranteed. In this specific case, pressure from environmental groups and gardeners has also led to better labeling of bulbs from the Netherlands, the source of most of the bulbs sold in the world, as "bulbs from wild source" or "grown from cultivated stock." Gardeners can be a powerful force for change!

Most wholesale nurseries and retail garden centers are conscientious about getting their inventories legally and responsibly, but this is not always the case. Always be aware of the supply chain and do not be afraid to ask questions. Palms, cycads, ferns, lilies, carnivorous plants—all are being depleted in the wild for horticultural use. In the United States, a number of species of rare cacti and orchids have been harvested, sometimes for personal use but often for resale, and more common ones have become rarer through overcollection. The National Park Service has started inserting microchips in some cactus plants so they can be tracked in the resale market as unscrupulous speculators often remove thousands of plants. Additionally, *Trillium* species are dug and resold throughout the United States, and rock garden enthusiasts have been guilty of "liberating" alpine plants. Some harvesters may practice some conservation measures, such as collecting only small numbers from each population, but many are out for quick money and take everything they encounter. Even selective harvesting may have long-lasting effects, because collectors generally take plants when they are in flower and easy to identify. That means the loss of not only these plants but also the ones that would have come from their seeds. And although one collector may be frugal, there may be others using the same site.

When more of a species is taken than can reproduce, the harvests become unsustainable. How do we know when that point has been reached? In general, we can't tell—the answer depends on many things that we don't know. How many populations of that species are there, and how many individuals are in those populations? At what age do they start flowering and fruiting, and how many seeds or spores does each mother plant produce? How difficult is it for seeds to germinate, and is predation of those seeds or seedlings high? It is difficult to answer these questions without detailed studies. In the absence of such studies and oversight by responsible agencies, all wild harvest should be considered unsustainable.

Sometimes permits are granted, usually for a fee, to go into public lands and harvest materials for private or commercial use. The permits set limits on how much can be harvested, calculated by land managers who believe they know what can be taken before harm is done. Unfortunately, many collectors prefer to harvest illegally, without limits. If you suspect that any material—plants, moss, floral greens, mushrooms, etc.—is wild-collected, ask to see a permit or ask the store that you are buying from to verify the supplier's permit. Plant groups to be especially wary of are cacti and other succulents, native wildflowers, bulbs, orchids, cycads, and carnivorous plants.

On some occasions, harvesting from the wild is not only appropriate but essential. As land is developed all over the world, plants and habitat are lost. County and city agencies and nonprofit land conservation organizations often organize plant rescues to go in ahead of the bulldozers and salvage whatever they can. These plants are then used in restoration projects in nearby places. Salvage should be considered only when destruction is imminent, not just planned. Your group can ask the developer for permission to dig up some of the plants before clearing begins.

Harvesting is more sustainable if done correctly. If your intention is to use the plants in a garden, then you should consider the genetic concerns about gene swamping discussed above. The rule of thumb is to collect just a few seeds from any given plant, taking from a number of mother plants throughout the population. This ensures that no particular genotype is depleted in the wild and that your collection will be genetically diverse. If there are desirable attributes in only one or a few individuals within a population, take just a few seeds and propagate later plants from them by seeds or cuttings from the first generation. You should never take more seed than is absolutely necessary, and any extra should be returned and scattered in the original population. Cuttings can also be taken, if appropriate for the species, following the same guidelines as for the seeds.

Before taking any wild material, you should research its conservation status. Is it on a federal, state, provincial, or other list as rare or threatened? You can check in libraries and on government web pages. If the species is considered rare or threatened, no material—seeds, bulbs, plants, anything—should be taken without legal permission and careful thought. You should be absolutely certain that you have correctly identified the species. Knowledgeable biologists must carefully orchestrate any recovery of rare species. Volunteering at botanic gardens, universities, and nonprofit organizations is the best way for the rest of us to assist in the recovery of rare species.

Depleting the Bounty of the Land

Garden plants are not the only plant products with a huge market. Many plants are stripped of their branches or flowers for the cut flower trade, affecting their biological communities. Other harvesting can do similar damage. I became aware of the problem of collecting nontimber forest products in the early 1980s as a greenhouse manager and later operations manager for an interior landscaping company, my first full-time job with plants. We would typically put the plants in their nursery pots inside decorative pots. To mask the "unsightly" gap between nursery and decorative pots, we would use a thick covering of moss, which is abundant in the forests of the cool, moist Pacific Northwest. Where did we get ours? From a collector known to me only as Tom. Tom was a character, missing a few teeth, always with a cigarette hanging out of his mouth, a little bit rough, but not in a really scary way. Tom would show up with a pickup truck full of fresh green moss in plastic bags with duct tape handles. Over time I became alarmed at the amount of moss Tom was bringing us, knowing that we were by no means his only customers and that there were other "Tom"s at work in the area. When I inquired about his collecting locations and methods, he informed me that there was lots of moss out there, something I imagine all the Toms believe to be true. How many of them can collect over how many years before the loss becomes critical and affects the local plant and animal communities? What other organisms depended on the harvested moss for habitat or food?

Do Native Plants Connect Us with Our Community?

The strongest benefit of using native species in gardens may be in providing a sense of place. Environmental psychologists use this phrase to mean that a landscape is unique. Growing native species connects us with the history of the land and ourselves. Wendell Berry, the great writer and advocate of sense of place, said it succinctly: "You can't know who you are until you know where you are." If we embrace the characteristics, including the plants, that make a place unique or special, we gain an attachment and a sense of belonging to that place. One of Berry's seventeen rules for developing a healthy, sustainable local community is that nature—the land, water, air, and native creatures—must be considered a member. Aldo Leopold said much the same thing in *A Sand County Almanac:* "The land ethic simply enlarges the boundaries of the community to include soils, waters, plants, and animals or, collectively, the land." When we connect with our environment, we enrich our lives.

Final Thoughts

In reading this chapter, you may come away believing that I do not support the use of native plants in the landscape. Nothing could be farther from the truth—I am always disappointed when I travel and see gardens made up of the same species, whether in London or Sydney. I look for that sense of place in the plant selections to tell me I am somewhere unique. But many people, in their zeal to use or encourage others to use native plants in landscaping, make many incorrect and even potentially hazardous assumptions. Native plants are wonderful choices for our landscapes, as are many nonnative species. The conscientious gardener must be informed and thoughtful, even about something that seems as obviously good as using native species in the garden.

Guidelines

- If there is a local native plant society, join it! You will learn about native plants through lectures and hikes. You will meet other people interested in native plants. You will have fun!
- Explore plants that are suited to your climate, soil, and other conditions with local garden center experts and in local botanic gardens. Native plants might be the answer to your landscaping questions, but they might not be.
- Observe and study how wildlife use plants in your area. Which species do they feed on, and where do they build their nests? Visit local libraries or join bird-watching groups such as the Audubon Society to find answers. Plant native species that are helpful to the native wildlife you want to attract.
- Know your neighborhood! Identify the native and especially the rare native plants near you in greenbelts and parks. Use only sustainably propagated local sources in your garden, especially if creating mass plantings or if the wild populations are small. Remember that your little patch of earth is connected to those around it and that genes travel farther than you may think.
- Ask the staff at your garden center where its plants are from. If they don't know, suggest that they ask their suppliers. Look for plants labeled as nursery-propagated or propagated from cultivated stock, especially if they were in one of the groups listed earlier. Check bulb and other catalogs for a statement about plant origins, but remember that even if it says they were grown in a nursery, that may have been between collection and sale.
- Volunteer to work with groups that salvage native plants. If you observe that an area with native plants is going to be developed, ask the developer if anyone will be salvaging them, such as a natural resources or parks department. If not, call those agencies—they may be unaware. If they are not interested, you might be able to take some plants with the developer's permission.

- Volunteer for native plant restoration projects. Parks departments, land conservancies, universities, botanic gardens, and many other land owners are undertaking such projects, which are great opportunities to learn about local native plant communities.
- Volunteer for projects to recover rare native plants. These may include collecting seeds for later use, monitoring rare plant populations, or helping to establish new populations. They may be overseen by botanic gardens; the Center for Plant Conservation, a network of conservation-oriented public gardens throughout the country; or researchers at universities.

CHAPTER FOUR

Aliens among Us

> There must have been plenty of them about, growing up quietly and inoffensively, with nobody taking any particular notice of them. . . . And so the one in our garden continued its growth peacefully, as did thousands like it in neglected spots all over the world. . . . It was some little time later that the first one picked up its roots and walked.
>
> —John Wyndham, *The Day of the Triffids*

In the spring of 1987 I went to Chile to do field work for my Master of Science degree. I was studying Winter's bark *(Drimys winteri),* a species in the Winteraceae, perhaps the most ancient family of flowering plants alive today. Winter's bark occurs in the middle of this long, narrow country all the way down to the very tip of South America. Its forms vary considerably, and I wanted to determine if the variation correlated with geography. Some of the funding for my trip came from the Washington Park Arboretum in Seattle, which wanted me to collect seeds of potential introductions to further its mission of introducing and displaying plants that are hardy in the maritime Pacific Northwest. The climates of Seattle and parts of Chile are remarkably similar, yet few species from Chile had been introduced into horticultural use in North America.

As I collected seeds, I started to worry about potential harm from the introductions. Many were colonizing species, many were dispersed by birds, and many produced lots of fruits and seeds. These were all traits that I had learned are associated with invasive species. Was I collecting species that would damage native communities in the Pacific Northwest?

I vowed to research the invasive potential of these species when I returned, using methods I assumed were developed and waiting for me in a paper in the library. I would not give the arboretum anything that seemed likely to become invasive. However, when I started looking in the library, I found that there was very little published on ecological invaders. Moreover, when the issue of predicting invasive potential was addressed, the biologists said it could not be done—invasions were too idiosyncratic, the chance interactions of the biology of the species and the plant and animal communities that existed before invasion. This made no sense to me: I had traveled enough to know that it is generally the same species, like oxeye daisy *(Leucanthemum vulgare)* and gorse *(Ulex europaeus),* that invade over and over again, while there are very popular species, such as boxwood *(Buxus sempervirens)* and many species of *Camellia* and *Rhododendron,* that are used all over the world and never invade. In other words, I had found the question that for more than twenty years has interested me most: why are some species invasive and others not, and how can we distinguish between them? We must consider this question when we use nonnative species in landscapes, to ensure the protection of habitat for natives.

What Are Invasive Species?

Nonnative invasive species are sometimes called *exotics* or *aliens.* Both of these terms mean that the plants are not native to the area, but since the words have other connotations, they are less specific than *nonnative.* Many people, including me, sometimes refer to invasive species as weeds. There are nearly as many definitions of *weed* as there are species of weeds, but the best one is "plants that have a negative impact on a desired state of the landscape." Any species, including native ones, can be considered weeds if they occur in places they are not wanted. It is therefore difficult to define *weed* in biological terms, because it is usually used to express a management preference. There is a legal term, *noxious weed,* for any plant that is injurious to humans or their interests, including livestock, freshwater systems, and natural areas. Formally calling something *noxious* often triggers regulatory actions, so this is usually reserved for only the most damaging plants, but sometimes even very damaging species are not listed as noxious. A classic example is kudzu *(Pueraria montana* var. *lobata),* an Asian vine that is now widespread in parts of North America. It covers forest trees, starving them of light, but it is not listed as a noxious weed by the United States government because at this point mandating control nationally is impossible. Some states where it is not widespread do have it on their noxious weed list. Kudzu has the potential to be invasive even in colder states and should *never* be planted.

Unlike weeds, invasive plants can be defined in biological terms. In 1997 the American Nursery and Landscape Association convened a group of biologists and horticulturists to discuss invasive plant issues. Among other things, the group developed a definition that works on many levels: invasive nonnative plants are those that can spread into native wilderness or managed ecosystems (everything from your backyard to agricultural settings), develop self-sustaining populations, and become dominant and/or disruptive to those systems. This definition has many strengths. By including potential spread, it captures our uncertainty about which species among all nonnatives may ultimately cause harm. It also notes that the populations must be self-sustaining—we are worried not about species that merely persist years after planting but about those that actively renew themselves, usually through seed reproduction. Finally, the definition recognizes that some species can be disruptive without being visually dominant, a sneaky group sometimes called ecosystem engineers. How they and other invasives launch their attack is described next.

Mounting the Invasion

Invasions occur in several stages. First, the plant arrives at a new location. Then it survives and begins to move beyond gardens and crop fields, into wildlands. Finally, it thrives and multiplies there, increasing in numbers of individuals and populations.

Plants are introduced to new areas in many ways. Like animals, they migrate—although usually much more slowly, as wind, water, and animals spread their seeds over long distances. Humans, however, have sped up this process, increasing the scale of its effects. Starting in the 1500s, trade goods were shipped between Europe and North America. Often, however, the goods that the Europeans brought were not equal in weight to what they picked up, so they loaded their hulls with extra soil to keep their ships weighted down while on the high seas. When they got to port they dumped the soil to make room for the trade goods, and many invasive plants escaped from these mounds. A paper by Viktor Mühlenbach discusses the role of ballast and the railroads that connect ports with other cities and allow the transport of invasive species from soil mounds to distant locations. Even the move in the early 1900s from dry to wet ballast did not solve the problem, as seawater, the new ballast of choice, has been the source of many aquatic invasions.

Humans have been responsible for many large-scale intentional introductions too. Gardeners have always sought out new species, and from the era when gardening was limited to the wealthy on vast estates, to today, when it has become a major hobby for millions of people, the interest in and demand for more species to grow and enjoy has increased exponentially. Starting in the late 1880s, European plant explorers forever changed the face of Western horticulture by climbing mountaintops in Asia and Africa to bring new species to the rest of the world. Some are legendary. Robert Fortune collected in Japan in the late 1800s, and numerous plants, such as the windmill palm *(Trachycarpus fortunei)*, have a species name in his honor. Joseph Rock, an Austrian who immigrated to the United States, collected extensively in China and Tibet between 1924 and 1949, when the political situation forced him to leave. Plant exploration enjoyed a new phase as China reopened to

Driving the Daisy

Automobiles can move seeds surprisingly well, as shown by Moritz von der Lippe when he was a graduate student in Berlin. He conducted his elegant research project in motorway tunnels a minimum of 700 meters long that connected Berlin with the surrounding forest. A high wall separated the lanes leading into the city from those leading out, so he could track whether the seeds came from inside or outside the city. Von der Lippe set seed traps about 150 meters from the entrance leading out of the city at ground level and slightly higher. The latter did not contain many seeds, proving wind dispersal was largely not responsible for the seeds' movement. He then removed and germinated all the seeds, totaling 11,818, and found that 204 species had been deposited in one year. Most of the species were not from near the tunnel, about half were not native to the Berlin area, and 39 are problematic in some parts of the world. While earlier studies had found that cars carry seeds, this was the first to demonstrate that they also deposit seeds. Therefore, even those of us in cities cannot be certain that seeds from our plants will not be distributed into wildlands. They *will* jump the garden gate, and we cannot control their movement. But we can do our best to keep our vehicles free of seeds and other plant material. Boats, camping equipment, and even shoes and clothes can transport seeds and plants, which should be disposed of in the trash rather than just knocked to the ground.

the West in the 1970s and collectors such as Dan Hinkley of Heronswood Nursery found plants there—and elsewhere—to enchant gardeners.

Most of these introductions have caused no harm and been valuable and important additions to food crops and landscape horticulture. Unfortunately, even imported seeds of noninvasive species are sometimes contaminated by invasives, so most countries now have laws that mandate inspection of imported seeds and set strict limits on the allowed amount of unintended seeds. However, even some intentional introductions have behaved badly: my research in the United States in 1997, and that of many people in other countries, shows that more than half of the invasive species found in developed countries around the world were introduced for the horticultural trade. Some percentage, however low, of introduced species will have invasive tendencies, and the sheer number introduced for gardening eclipses the amount introduced for other purposes. As our population grows, as more people become gardeners, and as nurseries feel pressure to introduce more species to engage their customers, we will spread more nonnatives than ever before. As it happens, many of the traits we prefer in garden plants are the same ones that increase invasive ability, such as fast growth, easy propagation from seed or vegetative parts (which makes them less expensive for growers to produce and therefore for us to buy), and high fruit production. Species also tend to go through intense periods of popularity, meaning more plants growing in more gardens, with a correspondingly greater probability of escaping by bird, wind, automobile tire, and so on. Invasion biologists refer to the phenomenon of increased use leading to increased invasion by many terms, such as *propagule pressure* and *infection pressure*.

When Aliens Attack

As gardeners we know that weeds compete with our garden plants for water, nutrients, and light and that lots of weeds means lots of competition. Some invasive species, however, have more insidious impacts, and we should worry most about them—the sneaky ones that biologists sometimes call ecosystem engineers. These species may change soil chemistry or water flow and use on a site, or they may be catalysts for long-term changes known as cascade effects because each one leads to another.

Nitrogen-fixing species are classic examples of ecosystem engineers. They occur in at least seven plant families, most notably the pea family. Nitrogen fixers have symbiotic bacteria in their root nodules that change atmospheric nitrogen into forms the plants can use, allowing them to move into areas whose top soil layers—which contain the most nutrients—have been removed. They also leach some nitrogen into the soil. This can be a good thing: many nitrogen fixers are early-successional species that, in their native ranges, ready the soil for later stages of plant colonization, and some people sow them in their vegetable gardens in winter to prepare for the growing season in spring. Problems arise, however, in places where the native soils, such as sandy soil, are naturally low in nitrogen and nitrogen fixers are not indigenous. The species that do best in these soils are often less competitive under higher nitrogen levels, and the increase in nitrogen may lead to invasion by other nonnative species that will crowd them out. Even when nitrogen fixers are removed, the soil's nitrogen level may stay sufficiently high that native species cannot recover without assistance—the ecosystem will have a different soil chemistry, which may last for several years. Soil chemistry may also be more altered in older stands than younger ones, and thus the area may be harder to restore.

Perhaps more problematic are the species that cause a cascade of changes to the ecosystem. Each ripple may appear slight, and, as with the ecosystem engineers, the effects may take time to develop and be difficult to reverse. For instance, in much of the world, species of *Fallopia* (sometimes called *Polygonum*) known commonly as knotweeds are serious pests in riparian systems. These species and their hybrids aggressively spread by rhizomes (underground stems) that break apart during floods and regenerate downstream. They grow in dense thickets and may totally replace native forest.

Lauren Urgenson, while a graduate student of mine, found that knotweed produces 70 percent less leaf litter than native trees in the Pacific Northwest, and its litter has a different ratio of carbon to nitrogen. Nitrogen is important to plant growth and is the central element in chlorophyll, the green pigment in leaves where photosynthesis happens. When leaves change color and die in the fall, it is because the plant is relocating a portion of its nitrogen and other nutrients for winter storage. Knotweed moves 76 percent of the nitrogen in its leaves to its rhizomes, compared to the 38 percent that willows and 58 percent that alders move to their roots. Many animals, especially insects, rely on the nitrogen in the leaves that fall to the ground, so the leaves' lower quality means less food for them. The loss of native trees that keep more nitrogen in their leaves cascades into a loss of insects, a loss of fish that eat the insects, and a loss of birds and other vertebrates that eat the fish.

There is no doubt that invasive species can do major damage, but it is difficult to measure this harm in monetary terms. Because the effects can be so wide-ranging, one would have to consider such factors as the cost of weed control, including herbicides and labor, in agricultural, wildland, and garden settings; lost crop yield for farmers; and, to be really complete, lost recreational areas, the tourism dollars that go with them, and lost fisheries. David Pimentel, a professor at Cornell University in New York, has made a couple of heroic efforts, most recently with colleagues in 2005. His conservative estimate takes into account mostly control costs because the other financial losses are much more difficult to calculate. His estimate is that in the United States alone, invasive plants cost a whopping $25 billion a year!

Signs of the Alien

It would be great if, as I thought when I was in Chile, it were easy to determine invasive ability. There are still enough unknown variables to make prediction an inexact science, but we have learned many things about nonnative invasions in the years since my trip to Chile, and we know that it is not as random as the papers I read then suggested. From the federal to the individual level, we have access to information that can help us make educated guesses.

One of the easiest and most accurate ways to determine a species' invasiveness potential is to see what it has done following introductions in other parts of the world. Today it is quite simple to use Internet search engines: simply type in the scientific name of the species and include the search terms *invasive* and *weed*. Check to see if the pages really say that the species is invasive—if they suggest the species as an alternative to an invasive, those terms will also appear. Consider the sources of the pages too—educational and government web pages are generally reliable, while the accuracy of others is highly variable.

Population biologists like to consider changes in population size using models of how many species are born or immigrate into a population versus how many die or emigrate out of a population. We can think of invasive ability along these lines too. The presence (or absence) of biological traits that increase invasive ability is a useful indicator, and I have found that some of the best predictors are traits that suggest high reproductive rates (many births) and good stress tolerance (fewer deaths). If the species has good dispersal mechanisms, such as tasty fruit easily recognizable as food to birds, its opportunities to move and establish new populations are also increased.

High reproductive traits include production of lots of seeds, although that becomes less important with efficient dispersal methods, such as bird consumption. High seed production is often, in fact, an indication of wind dispersal, a strategy of overwhelming the landscape with seeds in the hopes that some will manage to land in spots suitable for germination and growth. Even species that are not able to reproduce by seeds, perhaps because of a lack of pollinators, might use vegetative spread—reproduction by, for instance, rhizomes, stolons, soil and tip layering, or bulblets—to grow very fast. Such reproduction allows many new individuals to form, all genetically identical to the parent plant, which is established and adapted to the local growing conditions. Vegetative reproduction also helps plants deal with disturbance, a key stress. When a disturbance happens—whether natural, such as flooding, or human induced, such as plowing of soil—vegetative reproduction means the fragmented plant parts can regenerate quickly. Not all species can reproduce vegetatively, and invasive species are generally more capable of it than noninvasives. Invaders are also more likely to be able to regenerate the aboveground parts when a disturbance removes them, as long as some of the root system is intact.

Nitrogen fixation is another stress-tolerating trait, as it allows invasive species to colonize low-nutrient soils. Many invasive species are also semi-evergreen, meaning that they have photosynthetic stems or the ability to drop some leaves in response to cold or drought. Losing some leaves, which take water to maintain, can help a plant continue to photosynthesize at low levels until the stress has passed. Similarly, seeds that can lie dormant in the soil for many years before sprouting allow species to skip germination in years with poor environmental conditions.

Do I expect you to know all these traits for each new addition you consider for your garden? No, but keep them in mind as you select, grow, and observe plants. Your garden center should have information on these traits for each species it sells, and thinking about invasive ability in these terms will help you know which questions to ask. You can also consult the appendix for a list of garden species that are known to invade, their impacts, and where they should be avoided.

Opposing the Invasion

Gardeners, take action! Now that you know about invasive plants and what they do, you are ready to fight the aliens. Read on to plan your attack!

Spreading the Word

Most garden centers sell at least some species that invade wildlands in their areas. In some cases, they may be unaware of the invasive potential of a species, but often they will cite consumer demand as a reason for carrying them. Garden centers have a constant fear that if customers want something and cannot find it, they will simply go on to the next store. As you become a more conscientious gardener, make sure that your friends and neighbors benefit from your knowledge. Be sure to discuss with garden center managers, in a helpful and polite manner, your concerns about their selling species that invade nearby wildlands. Consumers have powers that they often do not use!

Like garden centers, the garden writers for your favorite publications may also promote invasive species. Again, they may be unaware of the invasive ability of the species, and it is helpful to politely let them know

What Is a Wildflower?

One area of particular concern is wildflower mixes, which often contain invasive species. Many consumers may think they are getting nothing but natives in these packages, some of which, to make things more confusing, are labeled "regional mixes." By this the seed companies mean that the flowers in the mix are suitable for growth in the region, not necessarily that they are native. In fact, nearly all wildflower mixes have at least some species that are not native to the region in which they are sold—the important thing, in the companies' eyes, is to provide an array of easy-to-grow flowers. Unfortunately, easy to grow usually also means easy germination, fast growth, lots of flower and fruit production, and stress resistance. Do these sound familiar? Rather than buy a mix, make your own with seeds of native or noninvasive nonnative species from your garden center.

for future articles: you can drop them a note about your interest in protecting wild areas. Garden writers who publish online should be aware that people far outside their region may read their work, so they should research whether a species is invasive in other areas.

Neighborhood Watch

Several years ago I was driving by a park near my house and glanced over to the side of the road. There was a small green plant with small white flowers growing all over the roadside. It looked like photos I had seen of garlic mustard *(Alliaria petiolata),* an invasive species that has caused many problems in forests in eastern North America but had not been reported in Washington. After verifying its identification, I posted a notice on the local native plant society email list. Someone responded to me, "Oh yes, I saw that population last year, but I just figured people knew about it." Well, no, they did not. Ultimately, it turned out that there was garlic mustard in a few city parks, possibly spread through contaminated maintenance equipment or shared mulch. If we had known this was happening a year earlier, could the spread have been halted?

When you see a plant that you think may be invasive—perhaps it is unfamiliar or spreading aggressively—do not be shy about letting people know about it. Do not assume that authorities know it is there. Your local government will likely have some department that deals with invasive plants and noxious weeds. Extension agents can also help get the word out to others. If you are not sure of what you found, they can help you figure out what it is. Local colleges and universities may have herbaria—collections of pressed and dried plant specimens—and be able to help you identify your plant and find the proper people to report it to. At the very least, you should post a notice to local native plant and gardening email lists. The point is to let others know in case they need to take action. Make a careful note of where you found the plant so you can give directions. You might also want to preserve a sample to show others by pressing and drying leaves and flowers or fruits inside a large book. The more people doing this sort of early detection, the likelier we will be to prevent some invasions in the future.

Organizing the Vanguard

It is important that gardeners take an active role in guarding against invasive plants, but it is not always possible for us to know which plants require extra caution. It would be easiest if aggressiveness in the garden were directly correlated to invasive ability in the wild. Some rampant garden plants do invade wild areas, but many have limited dispersal abilities and are unable to crowd out other plants outside the garden. If you have a plant that is spreading rapidly in the garden, check it out in the appendix and online. It may be all right to continue using it in the garden, but if it is really aggressive, do not share it with other gardeners—it may become invasive in the future, and even if it does not, your friends may not appreciate a garden invasion. Further confounding our ability to predict invasions is what biologists call the lag phase, the amount of time between introduction and invasion—sometimes decades. (The time span may be a reflection more of humans than of the plants: we are usually unable to detect an invasion until there are so many plants and populations that the fact becomes undeniable. For instance, in the early 1900s, an Australian extension agent noticed an invasive grass that had caused problems in other parts of the world and issued an alert to local landowners. Suddenly, the grass was everywhere! Or had it been present for some time and unnoticed because no one had bothered to look at yet another grass out in a pasture?)

There are several possible reasons for a lag phase. (The Jeffrey Crooks and Michael Soulé article listed in the Technical Papers for this chapter is a good overview of some.) One is the propagule pressure already discussed: a species might be introduced and used but not become popular for many years, or new techniques might allow it to be mass-produced when it was formerly difficult to propagate. An invasion might also lag behind introduction because birds, which often use visual cues to find food, do not recognize a new fruit as familiar enough to try eating it. Specialized pollinators, if there are any, might not have been introduced, preventing seed production. In one case, fig trees used as ornamentals in Florida did not invade for decades, until their specialized pollinator, a tiny wasp, was accidentally introduced.

Most gardeners do not have the resources to determine which plants are going through a lag phase before invading, so we would like to rely on an authority that can do this for us. The most obvious candidate is the national government, which regulates imports and exports and has access to people, money, and supplies for testing, so many people are surprised to find that most nations, including the United States and the European Union countries, do not assess invasive ability before a species is introduced. Australia and New Zealand have been formally assessing risk for several years through state and federal programs, and other countries, including the United States, are exploring whether they should, but in general most seeds and plants are only inspected to see if they have insects or diseases. The situation here may change in the next two or three years; to help things along, you can let your elected federal representatives know you support the United States Department of Agriculture assessing species for invasive potential before they are imported.

Until the United States follows Australia's and New Zealand's examples, gardeners will be, in many ways, the front line of defense, so do your part to win the war. Be wary when someone tells you that a cultivar of an invasive species will not be invasive. There are certainly species for which this is true, such as some variegated cultivars of English ivy, but often it is not. Ask for the rationale behind the claim. If the plant has not been seen in the wild, find out how long it has been available. If the period is less than ten years for a herbaceous plant or twenty-five for a woody species, be skeptical: it may be going through a lag phase. If the explanation is that the cultivar does not produce seeds, find out why. Some cultivars do not produce seeds under some circumstances but will under others. For instance, purple loosestrife *(Lythrum salicaria)* cultivars may not produce viable pollen, so seeds cannot form if only that cultivar is used. However, it may be able to produce seeds when pollinated by another cultivar in that species—which you or a neighbor might have.

Tending Your Garden

No matter how large or small your plot, you should be a good steward and develop that land ethic that Aldo Leopold urged us to embrace. Remove invasive plants that move into your garden and do not introduce any of your own. Invasive plants do not obey property lines and will eventually move outside your garden. Some gardeners will be tempted to bargain for their favorite species: as long as they prevent seed growth on their plants, it should be all right to grow them, right? But will they always be around to remove the seeds? What if they go on vacation? What if they sell the house and the new owner is less vigilant? There are so many fine alternatives to every invasive species, this risk is just not necessary.

While we may grieve if a favorite species becomes problematic and needs to be removed, equally wonderful species can replace it. First, identify what you like about the species. Is it the flower color? The long bloom time? The drought tolerance? Write down everything, then take that list to a garden center and ask the staff to help you find safe species that have the same wonderful attributes but are noninvasive. You do not need to restrict yourself to native species—most nonnatives are not invasive. You may not find all the attributes in one species, but usually it is not too hard to come close, and the new species may have terrific features the invasive one did not. Finding alternatives to invasives is something that many groups, such as the PlantRight program in California (see Resources), are working on right now. Your local weed district, extension agents, or a library with a good gardening section can help you locate previously developed lists.

Final Thoughts

We have come a long way since I returned from that trip to Chile—we know more about the impacts, spread, lag phase, and traits of invasive species, and we have gone from thinking that invasions cannot be predicted to actually predicting them (at least in Australia and New Zealand). If I were returning from Chile today, I would have the methods I developed for my PhD, as well as others, to evaluate the species I wanted to bring back. But in 1987, all I could do was make an educated guess, and while the arboretum got most of the seeds, it did not get them all.

In *A Sand County Almanac,* Aldo Leopold writes, "Industrial landowners and users, especially lumbermen and stockmen, are inclined to wail long and loudly about the extension of government ownership and regulation to land, but (with notable exceptions) they show little disposition to develop the only visible alternative: the voluntary practice of conservation on their lands." This is as true today as it was then, but how can we change the situation? Perhaps if we provide adequate information to help landowners take better care of their property, they will be more inclined to do so. Horticulturists generally welcome the idea, but do they have adequate tools to tackle one of the most important issues, preventing the introduction and spread of invasive species?

In 2001 I cochaired a workshop on horticulture and invasive plants at the Missouri Botanical Garden in St. Louis. We brought gardeners together with professionals from commercial nurseries, landscape architecture, and public gardens, as well as members of the federal and several state governments. We invited representatives from Australia, New Zealand, and Great Britain for their perspectives as well. Over three days we got to know one another and heard our respective positions. At the end of the three days we had all agreed to the same principles and findings. They were, essentially, that we all benefit from the richness of species that are introduced for many uses, but the current magnitude of introductions across the world by humans is unprecedented. We agreed that a small percentage of species cause harm and we should be pursuing research and education to prevent and detect invasions at the earliest stages. We also recognized that all of us have a role in achieving this.

Each profession, as well as the gardeners, also came up with its own voluntary Codes of Conduct, or best management procedures to prevent the introduction and spread of invasive plants. Each group had to present its codes to the other groups for comment but had the final say in what was included. Since 2001, the groups have been responsible for ensuring that those in their fields are aware of the codes and are implementing them. In the guidelines below, I summarize the codes of conduct for gardeners, as well as give some additional notes. The full text of the codes can be found at the website for the Center for Plant Conservation, listed in the Resources.

Guidelines

- Know which species invade in your region. This information is available from extension agents, native plant societies, government agencies that deal with invasive plants, libraries with good gardening sections, the Internet, and many other places.
- Be careful not to purchase species that are invasive in your area. Check the appendix for a list of species inappropriate for different areas and climates. When purchasing a species from a garden center, catalog, or Internet source, ask if it is invasive in your region. You may still need to do research to find out if that species invades anywhere else with a climate similar to yours. Ask gardeners who are familiar with the species if it has high-reproduction traits (vegetative reproduction, high seed pro-

duction, high germination rates) and high-stress-tolerance traits (semi-evergreen growth, nitrogen fixation). If it does, and if it has invaded in other places, choose another species instead.

- Do not take a lack of aggressiveness in your garden as an indication of lack of invasiveness in your region. Your garden conditions may be very different from those of wildlands, and the species may be going through a lag phase.
- Remove wildland invaders from your property. Even if you are in an urban area, studies have shown that seeds can move many miles on the wind or on cars.
- Seek safe alternatives to invasives. There are many lists being developed all over the world, so look for ones for your region from native plant societies, government agencies, and garden centers.
- Do not trade plants with other gardeners if you know they are invasive in your area or theirs. If a plant is multiplying rapidly in your garden, don't share it. Dispose of it appropriately, as discussed in the pest control chapter.
- When a garden center sells, a garden writer promotes, or a public garden displays invasive species in your area, politely bring the issue to their attention. Ask them to remove the plants or, in the case of garden centers, to stop ordering them for sale. You can also ask garden writers to further public education by writing about invasive species and their alternatives.
- If you spot a species that you think might be problematic, report it. Take it to extension agents, government agencies such as a department of agriculture, and/or herbaria.
- If you have the time, volunteer with parks departments, land conservancies, botanic gardens, and other places that are trying to remove invasive species from their land. You may learn some techniques you can use in your own garden.

CHAPTER FIVE

The Wild Kingdom

The swoop of a hawk . . . is perceived by one as the drama of evolution. To another it is only a threat to the full frying-pan.

—Aldo Leopold, *A Sand County Almanac*

Many books have been written about gardening to attract wildlife, and there are stores devoted to selling us bird feeders and houses, bat houses, and so on. But many products have been developed to repel, outwit, and even kill wildlife. What is this love/hate relationship with the animals in our midst? Ultimately, which side we are on comes down to our individual preferences and values, but even knowing our choice is no guarantee against feeling conflicted. People who are charmed by fluffy-tailed squirrels and put out nuts for them may also curse them for robbing the bird feeder.

I have always considered myself an animal lover. I have pets, I enjoy bird-watching, and I love to see animals in the wild when I travel. Nevertheless, I experienced some complicated emotions about animals when my husband and I moved into a new house in 2001. We loved all the birds that our large garden in a somewhat wild ravine attracted, but some of the animals? Not so much.

Several years earlier, a colleague had been the first to alert me to the unusual wildlife I could encounter as a gardener in the Pacific Northwest. She had just bought a new house and was extolling the virtues of the garden. "But," she sighed, "we have mountain beavers." I had no idea what she was talking about. Were they really beavers that lived in the mountains? What were they doing in lowland Seattle? Were they fictional creatures—a sort of Sasquatch? Was she pulling my leg?

The next time I heard about them was from the previous owner of my new house. She was telling me how much she would miss the garden, which I had seen when I viewed the house, although I had not had more than a quick look because it was overgrown with an impenetrable thicket of blackberries. "But," she sighed, "there are mountain beavers." Again I was confused. What were these creatures that caused gardeners to despair?

As I soon learned, mountain beavers are actually rodents, unchanged for fifty million years! (Another fun fact: the largest flea in the world lives exclusively on one subspecies in California.) Though they once also lived in Mongolia, they now occur only on the West Coast of North America, where they are broadly distributed in moist areas. As a biologist, I was intrigued by the idea of these ancient animals being a part of my landscape. So when the first thing I heard from my new neighbors after "Hello, nice to meet you" was various ways to trap and kill them, I was defensive on their behalf. I would be different. I would learn to live with these creatures.

Mountain beavers dig extensive tunnels with numerous openings, so it appears that many animals are at work when in fact there are usually only a few per acre. These openings were my initial chief source of concern: they are large enough to accommodate four-pound animals and are often in areas obscured by vegetation, making it easy for anyone to take a wrong step, fall into a hole, and break a leg.

Despite this worry, I was excited to learn more about my garden's inhabitants, and I watched them whenever I could. While they are largely nocturnal, mountain beavers occasionally venture out in the late afternoon. They timidly come to the mouth of a tunnel, sniff (they have very poor eyesight), then dart out, grab some vegetation, and run back. They also stack freshly cut plants just outside their holes for quick snacks, adding insult to injury when the treat is a prize hosta.

I was willing to step carefully to avoid holes, but my infatuation with coexisting with ancient animals came to an abrupt halt when they made a meal of newly planted mountain laurels. Still, I did not want them dead. I reasoned that their natural predators might be cougars or bobcats. I have three indoor cats—and I figured wild kitty urine smells about the same as domesticated. So I scooped the solid bits out of their nonclumping litter, carried the wet litter into the garden, and stuffed it down the mountain beavers' holes. This worked perfectly: I did not see them for weeks! After a while, however, they realized that no cougars would pounce on them and just started pushing the litter back out. But by then the image of my decapitated mountain laurels had faded, and I decided to learn to live with these animals again.

Then I discovered that overnight a favorite shrub—the rare *Neolitsea sericea* from Asia—had been chewed down to a nub. But I still didn't want the mountain beavers dead. I got a spray that smelled like rotten eggs to apply to plants you don't want eaten. That did not seem to keep the little guys away, though I certainly avoided those plants for days.

And so it goes. I vow to live with them, even love them as the almost-rare native animals they are. I give them apples, their favorite food, at Christmas. Then I discover that half of my beautiful small red-flowering Chinese dogwood *(Cornus kousa)* in full flower is gone and resolve to call the mammal curator of the University of Washington's natural history museum, who has told me that there is demand among world museums for mountain beaver specimens (the dead and stuffed kind) and has offered to trap some for that purpose. Come and get 'em! Then I take pity on them, until I find another favorite plant gone. Oh well, the plants usually grow back—eventually.

My dilemma is not unusual, although the specific animals may be. Wildlife control is a big business. Go online and you can find products and experts who will control your deer, rabbits, groundhogs, woodchucks, squirrels, chipmunks, moles, voles, bees, moose, raccoons, skunks, opossums, foxes, bats, bears, kangaroos, and yes, even birds. There is even a professional organization, the National Wildlife Control Operators Association, that trains and certifies people in corralling pest animals.

Making Wildlife Feel at Home

Despite some trials with wildlife in the garden, such as mine with mountain beavers, most people enjoy watching animals at home and abroad. The Fish and Wildlife Service (FWS) estimated that in 2006 in the United States alone seventy-one million adults viewed, fed, or photographed wild animals. This is good for the economy—in "Wildlife Watching Trends: 1991–2006," the FWS also calculated that $45.7 billion was spent on trips or equipment for this purpose in 2006. The agency's "Wildlife Watching in the United States: The Economic Impact on National and State Economics in 2006" report found that wildlife viewing in that year created more than one million jobs, with all those workers earning about $40.5 billion total. The FWS claims that state and federal governments benefit too—they get about $18.2 billion per year in related sales and income taxes, some of which should make its way back to managing wildlife.

If you want to bring wildlife to your home, you can take several easy steps to attract more, starting with a consideration of the land's carrying capacity. (If you prefer not to have animals in your garden, skip ahead to "Keeping Wildlife Out," near the end of the chapter.) Each plot can support just so many organisms at a basic level, so if you want more wildlife than what your land can currently provide for, you must identify what the wildlife needs and then supply it. Fortunately, this should be just a few basic necessities.

Food

First, provide something for them to eat. Research the animals you want to attract and offer them a buffet of their favorite choices. Plant fruiting trees and shrubs, but note that birds spread many invasive plants by consuming the fruits. Ask at the garden center and your local extension office to make sure you are choosing noninvasive plants. Native plants are a great way to attract native wildlife, so make those species a top pick. And keep in mind that birds and other wildlife do not eat only brightly colored fruit—they like seeds in general. Pines, firs, and other conifers can provide protein and fat-rich seeds, as can sunflowers and some grasses. Many grasses also provide leaves that are a good food source for many animals. Let some flowers go to seed and you may find more wildlife in your garden.

Water

Like humans, wild animals require water. If you are lucky enough to live near a natural source of water, you need do no more than protect it (see the chapter on water). If you do not, you might want to provide a birdbath, small pond, or other water feature. You will find that you attract not only birds, but squirrels, butterflies, bees, and much more. My husband, Brian, and I still laugh about the raccoons that would come to the small pond we built in our former garden. One time a group of three saw us watching them from the deck above and got on their hind legs and did a little dance—really!

Birdbaths will attract more than their namesakes: butterflies and even dragonflies and damselflies will also use them. A small, light birdbath or saucer might be just right for a balcony or deck, whereas a larger garden could support a formal birdbath or even a fountain. Make sure the bath has a gradual slope, a rough surface for bird feet to hold on to, and at least a section no more than three inches deep so little birds can use it. A small birdbath a few feet wide is fairly easy to maintain and should be drained, cleaned of algae, and refilled at least weekly to keep it healthy and free of mosquito larvae. Most mosquitoes take about two weeks to grow from larva to adult, so as long as you clean frequently you will not breed them.

Many birds and butterflies are attracted to drops or fine mists of water, and wildlife stores carry several systems to provide this in your garden. Drippers slowly release water, attracting birds with their sound. Misters can be attached to a garden hose to create a fine spray that birds enjoy flying through. Ken Jacobsen, a former state senator in Washington and a noted bird enthusiast, tells a funny story about these devices. Noticing a birdbath in one garden while campaigning door to door, he suggested to the elderly homeowner that she needed a mister. She indignantly replied, "I have outlived two misters, thank you! I don't need another!" There is no word on whether he got her vote.

Water gardening is an increasingly popular type of horticulture, as people realize how soothing it is to have water nearby and enjoy the wildlife it attracts. Simple ponds can be put together in a weekend with heavy plastic liners made for this purpose. Even if you have a small place, you can set pots of water plants in a relatively shallow ceramic container with no drainage, maybe add a recirculating pump for a waterfall, and create an entire contained ecosystem.

Along with the joys it brings, water means work. The type of water garden you build will dictate the maintenance you will need to perform, so consider this in the earliest stages of your design. Do not worry about keeping the water pristine: some decaying matter is all right and adds nutrients. But if you smell rotten eggs (sulfur) when you stir the water, it has too much decaying material and should be cleaned. A small pond or containerized water garden will need cleaning once a month, depending on its temperature, size, and water circulation—the light and warmth of the sun encourage algae to grow, but they may reproduce more slowly where the water is kept moving. Snails may also help remove algae, so you will have to clean less often. There are commercial animal-safe products available to prevent mosquitoes in ponds—check with your pond supplier. Garden centers can help you develop a suitable pond or water feature and set up a maintenance schedule.

The bigger the water feature, the more responsibilities you will have. If you want to build a large pond, you should contact the local planning office to determine if you need a permit and your insurance company to see what safety measures they recommend. If you add fish, such as koi, you may attract wildlife you did not expect. Raccoons feast on fish, and birds such as herons will spot them as they fly overhead. A grid of strings over the pond will discourage birds from swooping down, but it is not terribly attractive. Deep areas may allow the fish to hide out of the reach of other animals, but there is no way to fully protect them. It is probably best to embrace these natural feedings and not get too attached to your fish, but few people seem to feel that way. When I told a friend whose pet koi were killed by an otter that rather than mourn them he should celebrate that his garden attracted an animal as interesting as an otter, he looked doubtful.

Shelter

Animals need shelter to hide from predators, avoid temperature extremes, and raise their families. But they do not all like the same thing, so the more types of habitats you provide, the more types of animals will live in your garden. It is that simple, and fortunately, even a small lot can have a diversity of plant types and heights. You will probably also want spaces for humans to hang out and observe the animals, but remember that most consider us predators. Locate patios and observation spots where you can see your garden's wildlife but far enough away that they feel safe.

Many people put out readymade objects, such as feeders and houses, but even without these, you can offer shelter for a wide array of creatures capable of finding or making their own home. Raccoons and other animals love nesting under porches, so if you crave seeing them up close, you might provide a small out-of-the-way entry for them. However, you may find their scratching noises distracting, and you should think twice if there are skunks in your area, since they love getting under porches too, something most of us would prefer not happening. To provide a good alternative, create a brush pile of cut debris of various sizes on a base of rocks. (It should not be close to the house, because it may attract rats.) An alternative is a rock pile, with varying sizes of rocks and entrances that allow animals to dart inside.

Diverse canopy layers will also increase animal habitation. Look around your garden to identify the layers you currently have, then think about how you can add more structural diversity to attract more wildlife. Depending on space, you might want to have an open area to support animals that like warmer temperatures. It might have grasses that provide seeds for birds, as well as various low-growing perennials to hide small animals. Try leaving some grasses long and mowing others (after checking for nests), to present two types of grassland. A shrub layer can also offer shelter and nest sites, although not everyone can stand a dense tangle in their garden. Plants like wild roses are great because birds can get in but the thorns will keep larger predators out. Birds and other small animals will also use trees as safe perching and nesting sites.

If you need to have a tree removed, consider creating a snag. Most arborists will leave a section of the trunk standing safely and will even make a realistic torn-looking top if asked. The jagged edge will encourage the inner parts of the tree to rot faster by facilitating the entry of wood-decaying fungi. The snag will also be colonized by insects that help break down wood and which will be eagerly snapped up by a variety of animals, like woodpeckers. The birds' excavation of insects will open a cavity into the tree's rotting center, and soon other species will move in to raise their young in the hole. A whole ecosystem in one dead tree!

Safety and Sanctuary

Food, water, and shelter will bring animals to your garden, but what else will they find when they get there? We need to think about the mixed messages we are sending to wildlife, like my alternating between giving mountain beavers apples at Christmas and then trying to drive them away the rest of the year. Even the latter treatment is benign compared to something many of us do without a second thought: lovingly harbor predators in our homes. If you are trying to attract wildlife to your garden, especially small animals like birds, it is cruel to allow your pet cats to stalk and catch them at the feeder or bath. Putting a collar with a bell on your cat will not work because animals do not necessarily associate the sound with danger. It may seem against a cat's nature to stay inside, but with a few provisions like structures for climbing and pots of wheat grass to nibble, an indoor cat will live a happy life—and so, maybe, will your garden's birds. Keep in mind that your cats also face danger outside: many critters, such as the coyotes I find in my garden, would love to munch on them, and far too many pets die from being hit by vehicles. Indoor cats almost always live longer than ones that spend time outside.

Another aspect of leaving wildlife in peace is to never bring them into your garden yourself. Remember that many animal species mate for life and

Where's My Mommy?

If you find a young animal seemingly alone, do not pick up or approach it. Mothers of some species leave their young while they forage for food, and if they find that you have interfered with their offspring, larger mothers may attack you and smaller ones may abandon them. Unless you know the mother is dead, watch for a while to see if she returns, keeping an eye on the young animal to protect it from predators, including your cats. If you are really sure that something has happened to the mother, call a local wildlife shelter or animal control agency for advice.

have young to care for. Do not be a home wrecker by splitting them up, or worse, moving them to conditions that could kill them. When I was young, I brought home a box turtle from the summer nature camp I attended and left it in my neighbors' swimming pool area, which was surrounded by trees and a fence. After a couple of years it disappeared, and we all thought it had escaped or fallen to predators. About five years later we found a box turtle in the pool area shrubs. Was it the same turtle? Probably—the many trees and shrubs provided good cover to protect him from both predators and our eyes. Although he survived, he led a lonely life, and I regret that. Despite the fun of having and refinding the turtle, it is important to note that transporting wildlife like I did is a bad idea and, in some places, illegal. If the conditions are right for them, wild animals will find your garden. Gardeners are a patient breed—we are used to waiting for plants to grow and flower, after all. So give the wildlife and yourself time and perhaps check with others who have been successful in attracting the animals you desire.

If you do enough of the things recommended here, you might qualify to register your garden as a Backyard Wildlife Sanctuary with the National Wildlife Federation (see Resources for the website). A few states will also certify your garden as habitat. You do not need a large garden to do this—a plot of any size can be certified if you follow the required steps.

But why stop there? Wouldn't it be better to get your neighbors to also certify their gardens? More certified gardens means even more places for animals to eat and live—and if the gardens are close together, they might attract animals that need larger ranges. Wildlife-suitable gardens in urban areas can also serve as greenways between parks. Paul Beier and Reed Noss found that habitat corridors in otherwise patchy mosaics of landscape allow many kinds of wildlife to move between larger sites. If your neighborhood offers a series of wildlife-friendly gardens to move through, it will almost certainly attract increased numbers and types of animals.

Val Schroeder, from the small Washington State community of Camano Island, made that happen in her neighborhood. She developed and certified her small garden as a wildlife habitat, with food, water, and shelter. Then she started talking to her neighbors about it, and they got excited too. Soon the community was more than just lawns and foundation plantings: beautiful and diverse gardens now beckon to birds and other animals. Val's goal is to get more than one thousand of her neighbors certified, and she is nearly there.

Types of Wildlife

Birds

The most common type of wildlife most people want to attract to their garden is dinosaurs. Well, their descendants anyway. The debate about the exact lineage goes on, but most scientists believe that the suborder of dinosaurs known as theropods were the ancestors of modern birds. Fossil bones, especially some recently discovered in China, and bones of existing bird species provide much of the evidence: they have a number of similarities, including a fused collarbone and feet that have four toes, three pointing forward and one back. Feathers have also been found with some theropod fossils. The hypothesized ancestors of our birds were probably about the size of a crow and diverged from other dinosaurs about 145 million years ago.

What can you do to make these latter-day dinosaurs feel at home? Some birds that like to nest in tree cavities will enjoy a house positioned away from predators and where it is warm but not too hot. Just as with humans, the size of the house is more important than the cute features it might have. Small birds generally like houses with a small opening and no outside perch so predators cannot get in, and inside dimensions of six to eight inches high by four to five inches wide and deep. Bigger birds need bigger houses—check with your garden center or wildlife store about appropriate sizes for the birds you are likely to attract.

Even if you do not provide a house, you can help the birds build their own. In the spring, when many birds are building their nests, it can be fun to leave two- to five-inch lengths of yarn and natural materials like twigs or vines for them. You can watch them find the treasure, and you may even see it in a nest in your garden. Place your offerings near a feeder or somewhere easily seen—open places away from shrubs where predators might lurk and high up in a shrub or tree are good locations. Some birds also use mud in creating their nests, so you can fill shallow containers with it for them, a task kids of any age might enjoy.

The best way to attract birds is to provide the specific food sources they prefer. Some like to pluck small red fruits from branches, some like to hang on the branch next to a pine cone and peck out the seeds, and some like to forage in soil for insects. Providing a rich array of fruits and seeds throughout the seasons will nearly guarantee that birds will visit your garden, as will minimizing or eliminating the use of insecticides. Most insects are beneficial to the garden anyway, and many birds rely on them for food.

Plants offer fruits and seeds, but usually not year-round, so some people enjoy using bird feeders to attract birds. (In most cases, it is not necessary to use a bird feeder in the summer because there are ample insects and fruits available.) There are special mixtures of seeds and other foods to appeal to specific birds, as well as bird feeders for particular species. Bird specialty stores have a bewildering array but can help you make good selections for the birds you hope to attract. But be aware that bird feeders also appeal to rodents, so position them so you can enjoy the birds but not give rats the idea that your house is a great place to live. Clean up under them regularly, and keep the food in tightly sealed containers until you use it—if you do not, you may discover unwanted wildlife entering your house to find it. Also be aware that feeders may attract predatory birds like hawks, which will take advantage of such a concentrated source of dinner material. Seeing hawks is a big thrill, but knowing that you lured an innocent bird to its death is not. And, of course, cats are known to enjoy hunting at bird feeders. Do not place the feeders too close to vegetation that allows a hiding spot for such predators. Generally, putting the feeder about ten feet from some shrubs or trees prevents ground-based predators from hiding too close but gives the smaller birds somewhere to fly if they feel threatened. In some areas bird feeders may also attract wildlife dangerous to humans, such as bears, so do not use them if you live in such a location.

Feeders can be a major source of mortality for birds not only because of predators but also because they may help spread disease or hold contaminated food. Be sure to clean them regularly with hot water and allow them to dry before refilling with fresh, nonmoldy seeds. Some people like to use a solution of nine parts water to one part bleach to clean them.

If you want to attract hummingbirds, use one of their special feeders, with a mix of one part white table sugar to four parts boiling water, stirred until the sugar is thoroughly dissolved. Never use honey or sugar substitutes—honey can cause health problems, and the hummers need calories, so no diet sweeteners! Clean the feeder every week or so, getting rid of any leftover solution. Hummingbirds are attracted by red, but you should use a red feeder or one with red around the feeding hole instead of coloring the solution with food dye. Consider a feeder that has a place for the hummers to perch—it is true they can hover while feeding, but they enjoy a nice rest while eating just as we do, and this will help you get a better look at them. Some species tend to be territorial with food, so providing many sources will allow more hummingbirds to enjoy your garden. Finally, many, but not all, hummingbirds migrate to warmer locations in the fall. Check with wildlife authorities, specialist stores, and local bird guides to determine whether you should take your feeders down or keep them up through the colder months. If it is really cold and the solution starts freezing, putting it somewhere that will get some reradiated heat may help, or you may want to swap warm feeders from inside your house with cold ones from outside.

In many climates, you can also grow flowers to attract hummingbirds for much of the year, giving them a natural source of nectar. These flowers tend to be red, orange, or purple and have narrow tubes the birds probe for nectar. They often extend downward so the birds can hover beneath them. However, hummingbirds will visit nearly any nectar-rich species, especially in winter, when flowers are scarce. Some of the winter-flowering mahonias, which have yellow blossoms, are among their top choices.

Many other birds are migratory too, going south to warmer locations in the winter, then back north for breeding in the spring. Migratory birds have been suffering a decline in numbers, and most scientists have focused on declining habitat conditions, especially in the warmer tropical countries, as the main cause. Recently, however, more attention has been given to the habitats along their migration routes. As we fragment the land, it is becoming harder for them to find places to eat and rest. In the fall and spring it is especially important for those of us living between their destinations to provide food and shelter for these weary travelers. In addition to the good deed, our reward is sometimes seeing birds that do not live in our area.

Will feeding migratory birds in the fall and winter discourage them from flying south, increasing the likelihood of their freezing? According to ornithologists, while some waterfowl may not migrate in the fall if given additional food, this is not true of terrestrial birds. Therefore, you may feed them throughout the winter, but if you are concerned, you might check with local birding organizations.

Butterflies

Insects can be desirable wildlife too: for instance, many people enjoy having butterflies in their garden. Their larvae (caterpillars) often need to eat specific plant species, which extension agents and libraries can help you identify. Remember that those caterpillars chewing your leaves may soon turn into beautiful butterflies or moths. And once they reach adulthood, they will sip the nectar of many species, pollinating the flower in the process. Garden centers can help you choose plants to attract adult butterflies and often have special displays showcasing them. You can also appeal to butterflies with small amounts of fermenting fruits mashed up with sugar and placed around the garden. These might attract less-desired wildlife too, however. Some wildlife experts also suggest providing mud puddles so the butterflies can add minerals to their diet. In general, butterflies like open, sunny areas that are not too windy—ideally with some rocks that will heat up in the daytime, since they need to bask each day in a warm place. They are not afraid of humans, so you can put your butterfly garden near a patio or deck where you can sit and watch them. Cats, however, will want to catch them, so keep your pets in mind as you design your garden.

Bees and Other Pollinators

Butterflies are not the only insects that you should try to attract. Bees need a home too! African, European, Asian, and South and North American countries are all reporting decreases in pollinators, due to many causes, including loss of habitat through deforestation and development, increasing use of insecticides, and new diseases. European honeybees, in particular, often introduced into an area because they are great pollinators for a wide variety of plants, are being attacked by an array of pests, diseases, and mysterious "colony collapses."

Pollinator loss has many consequences. It can be a problem for the food supply, since many important fruits, such as apples, cherries, and pears, require cross-pollination before they can set. In 1979, K.T. Harper found that species with special requirements for pollination, such as an unusual flower shape, have the greatest risk of becoming rare or even endangered because they often have a limited pool of pollinators to begin with. Even common species with ordinary flowers are impacted, as the competition for few pollinators gets more intense. Natives frequently lose this battle to invasive plant species, which often entice bees and other pollinators into preferentially visiting them by providing copious nectar or protein-rich pollen.

Planting a diverse garden with lots of flower types will attract many pollinators to your yard and help support their numbers. Bees are attracted to blue and yellow flowers and generally visit a wide variety of flower shapes.

(They can also see ultraviolet patterns that we cannot.) Fragrance may entice them. Butterflies are day feeders and prefer brightly colored flowers, while moths feed at night and usually are attracted to flowers that are white or light-colored—easy to see in the dark—and open and fragrant at night. They usually have long tongues to reach into the corolla tubes of the flowers they favor.

To help protect your pollinators, limit the use of insecticides in your garden. Try other methods of pest control first. If you must use something that kills insects in general, apply it very carefully. Keep in mind that many species of native bees are ground-nesting, so try not to disturb nests if you find them (but note that yellow jackets and hornets, not usually pollinators, also sometimes nest in the ground).

A healthy pollinator population can help your garden in many ways. Having abundant pollinators will likely mean more fruit for birds and other animals. The small animals that come to eat the fruit might attract predators in turn. The tiny pollinator can be the key to a full wildlife ecosystem!

Bats

Popular culture often portrays bats as scary creatures that might suck your blood and turn you into a vampire. There is a real danger that in some areas they may carry rabies, but deaths from rabies-infected bat bites are extremely rare. And in fact, most bats are quite beautiful in their way. They may have faces only a gargoyle could love, but they have soft fur and, like humans, nurse their young (as do all mammals). Best of all, an insectivorous bat can eat tons of insects each year. They love mosquitoes—some species can devour up to six hundred per hour—which not only cause itchy bumps on us but also carry West Nile virus, malaria, and other diseases. It is wonderful to watch them hunt, swooping down gracefully to snag the next meal. They will not be caught in your hair: their echolocation system is far more sophisticated than what humans have developed, and if they can use it to

find a tiny insect, they most certainly can sense our giant human heads and maneuver around us.

Are you sold on bats yet? Then you may need a bat house: a warm, safe place where these nocturnal creatures can hang out (literally—they roost upside down) all day. Bat houses can be bought in wildlife specialty stores, garden centers, and many similar places. There are also simple plans available on the Internet. A bat house can provide shelter for up to twenty bats. It is best placed where it will warm up during the day, on a pole or building rather than in a tree—bats fly straight down when exiting their roosts, and branches will get in the way. Not all species of bats use houses, so check with wildlife specialty stores in your area to see if you need one.

Bats prefer to be about a quarter mile from a water source, such as a river, pond, or lake—even a swimming pool might do. They not only drink the water as they fly by but also feast on the many insects it attracts. If there are bats in your area, your bat house may soon attract some of these fascinating creatures, and you can sit in your garden in the evening free from mosquitoes, watching them hunt.

Herpetiles

The group of mostly small animals known as herpetiles, or herps, includes amphibians and reptiles. Frogs, toads, salamanders, and newts are called amphibians because they need water and land to complete their life cycle (*ampho* is Greek for "both"). If you live near a pond or stream, you can create hospitable habitats to help nurture our globally threatened amphibians. More information about the problems they are having can be found in chapter 2.

Consider leaving dead trunks to decay within several yards of a water source to provide homes for salamanders and other amphibians. Water features may attract them, though they are more commonly found near natural water. Amphibians regulate their body temperature through the environment, so when it is cool or cold outside, they will need a place to get warm.

This can be a log or other basking spot in a pond, or rocks or concrete that heat in the sun and slowly release their energy.

Despite knowing that there is a creek at the bottom of my hill, I had never considered my garden to be amphibian habitat. Then one day, weeding on my hands and knees, I literally came face to face with a redbacked salamander. I sat back and considered it as it considered me. I apologized for disturbing it and moved away. This is not a rare species, but it is an honor to have it consider my garden its home. Now when I work near the creek I try to move carefully to prevent harm to my small neighbors. I leave logs and stumps where they are as possible shelter and let leaves remain on the ground to provide burrowing spots. While I rarely see the salamanders, I like knowing that I am helping them by preserving their habitat.

Though most gardens are too small and contain too many dangers—such as cats, dogs, and pesticides—and not enough water to sustain most herps, if conditions are right you might try attracting toads with a toad abode, or small house for these slug- and mosquito-eating friends. Toad abodes are easy to create with old clay flowerpots—just score a two-inch opening at the rim, knock it out, and put the pot upside-down in the garden near water. You can also purchase any number of fun abodes in some garden centers or over the Internet. In cold months, however, toads go into hibernation, and if you have them, you may need a special structure, which is also easy to construct. Bury a twelve-to-fourteen-inch-long by two-to-three-inch-wide pipe or drain tile at a 30 degree angle, leaving about five inches above ground. Fill the bottom half with sand and the rest with decaying leaves. In the winter, toads will burrow into the underground part to stay warm, and they can live in the top part for the rest of the year, so you can make this your only toad abode. For a more natural look, leave piles of stones, woody debris, and autumn leaves around wet areas for use by toads and other amphibians.

Closely related to the amphibians are the reptiles, such as snakes, lizards, and turtles. Reptiles are less dependent on water habitats, but many need or prefer to be close to water bodies. They also require places to warm up, such as rocks and other basking spots. Many people have a fear of snakes, though most snakes are not venomous and may reduce some pest populations. It is a good idea to be cautious around them, however, unless you are sure of the species and know it is not poisonous. There are nearly thirty-eight hundred species of lizards, from tiny to quite large animals. Most are not harmful to humans, and they are often kept as pets. Turtles can also be great fun to have in the garden. They often live for many decades—some species can live more than one hundred years—and many will help control pests such as slugs and snails. They generally need a good water source.

Keeping Wildlife Out

What if you are on the "Enough with the wildlife!" side of the issue and want to discourage critters from entering your garden? Experts on wildlife control offer many suggestions, which are essentially the opposite of those suggested to attract wildlife. First, limit or remove any food sources. Although butterflies may like them, pick up fruits that fall from your trees, or they may attract all manner of wildlife, sometimes with unforeseen unpleasant consequences. (Fruit you leave out for butterflies will also attract other animals, but you may decide that is worth it.) A friend with a plum tree tells me that after opossums consume the downed fruits she finds evidence throughout her garden of their loosened bowels (which plums also cause in humans)—not a nice surprise. If a plant itself is the attractant, consider removing it, or at least not adding any more. In my case, that has meant identifying which species mountain beavers love—hostas—and not planting them in favor of species the animals seem to dislike—conifers.

Specially formulated pepper concoctions—commercial and homemade—sprayed on or around foliage may deter insects, rabbits, mountain beavers, and deer from munching on favorite ornamentals. Coyote or fox urine (you will want to acquire this commercially!) may help deter mammalian predators like rabbits and deer. A well-designed fence can keep deer from grazing on vegetables or prized ornamentals. If deer are unavoidable in your area and you don't want to build a fence, plant one or more of the many species they dislike.

Cover garbage containers tightly, and do not put fruit or vegetable trimmings in open compost containers. Never put pet food outside unless you want to attract wildlife in addition to feeding your pets. If you put it out, they will come! Also stay alert if you leave a door open near pet food dishes. Brian was astonished one day when he found a raccoon that had wandered through the open kitchen door at our old house in a densely urban area happily eating cat food.

Be careful even when intentionally feeding wildlife. One gardener gave some dog food to three cute raccoons that had come around. Soon there were nine, then fifteen, twenty-five—and now he has fifty-three raccoons visiting nightly for the feeding. One or two had been cute, but a veritable army of raccoons at his door was frightening—and he was afraid of what would happen if he stopped feeding them. Feeding wildlife in this way is also bad for them—once animals like raccoons become dependent on a food source, they tend to not forage for themselves as much. Since these artificially large populations cannot find enough food on their own, they also have an incentive to stay in one place rather than expand into new territory. Greater population density can then lead to aggressiveness among the animals and the spread of diseases.

To keep animals away, you also need to reduce their access to shelter. Again, do the opposite of what was recommended to attract animals: keep the shrubs in your garden trimmed and do not create piles of brush or rocks. Some animals may be scared away with motion-activated lights or sprinklers. Deer, a pest in many gardens, especially do not like sprinklers, which can be an effective and harmless way to chase them away. Do not forget to animal-proof your house. Put screens over vents and other openings to keep wildlife out of your crawlspaces and other areas. If you think that animals are in an attic or under a porch, try to identify all exit points and seal up all

but one. Use lights, loud noises, and other irritations to drive them out, then seal up the remaining opening. (If you do not want to do this job yourself, remember that every community has professional pest control operators.) You might also want to install a chimney cap to ensure that birds, bats, raccoons, and other animals don't move in. This will also mean that you do not have to worry about a blocked chimney causing a house fire.

Some of my fellow mountain beaver landlords have taken to evicting their tenants by trapping them and releasing them in other locations. If you are tempted to do this with a pest animal, think carefully about it. Many do not survive the relocation—if the new habitat is suitable, it will likely have others of the same species, and they will know better than the relocated animal where to find food and shelter. Even if you trap every animal on your property, animals from adjacent properties will just move into the newly available territory.

You should also be aware of the laws regarding how animals are handled. Before trapping, relocating, killing, or even using products such as bleach or mothballs to repel wildlife, check with local animal control officials to find out what laws are in force, even if you hire a wildlife control operator. While most operators go to great lengths to follow all laws, some will assure you that their proposed plan is legal without checking whether this claim is true.

Final Thoughts

I once watched a national gardening show that featured a couple who live in my area designing their garden to attract mountain beavers. I was astounded, but maybe they were right. Although I may curse my mountain beavers, whenever I catch sight of one I am awed and sit barely breathing so I won't frighten it away. For fifty million years they have been burrowing, raising young, and yes, eating plants quite successfully. Even when I spot a favorite plant sheared and stacked up near a tunnel opening, I secretly admire them.

Having wildlife in our midst is a gift. Hummingbirds zooming around with confidence at incredible speeds never cease to amaze me. Baby raccoons playing chase and tussling or Steller's jays squawking as they pursue the triumphant bird that captured the peanut a neighbor left for them entertain me. The bees that pollinate my apple tree help to feed me. These animals belong here as much as, if not more than, I do. While I might not want a squirrel to nest in the attic, watching him bury sunflower seeds he harvested from my plants, then seeing a sunflower grow from them the following year, enriches my life. An insect eats a plant and is eaten by a robin, which is then eaten by a hawk, and we reflect on the cycle of life. All we need to do is invite them in, nurture them, and quietly watch.

Most of us live in urban areas and are increasingly out of touch with nature, but even in urban areas, wildlife abounds. Pigeons are in nearly every urban area around the world. In some cities endangered peregrine falcons learn to nest on skyscraper ledges, which substitute for the cliffs they formerly used. Raccoons, naturally woodland creatures, are very smart and thrive in urban areas, where they have learned to live with humans even when rarely seen. We can learn adaptability and resilience from them.

In an urban area, a garden may be an island of wildlife heaven, surrounded by a sea of wildlife hell (hot pavement, no shelter or food, speeding vehicles). If, unlike Val Schroeder, you are unable to persuade your neighbors to make their property more wildlife-friendly, do not despair. Inevitably, some birds and other creatures will find your oasis. They might not be the ones you first hoped for, but enjoy their songs and antics, and you will be enriched.

Guidelines

- Learn about wildlife native to your area by joining local nature clubs, consulting area guidebooks, and visiting wildlife specialty stores. Learn which animals are good garden inhabitants and develop a plan to make your garden attractive to them.

- Increase food sources so a few more individuals in the desired species can visit your garden. From wildlife specialty stores and libraries, find out what the animals you want to attract eat. Favor natural sources, including native plants. If you provide special foods in feeders, keep them clean and filled and protect the feeding animals from predators.
- Add one or more water features to your garden. Small gardens can have small birdbaths, and larger gardens can have ponds that will attract a wider array of animals. They should be developed with maintenance in mind: do not build a feature larger than you are willing to clean.
- Add structural diversity to the garden. Provide many layers for nesting, shelter, and foraging, from open grasslands to taller grasslands, shrubs, and multiple layers of trees.
- Include pollinators in your definition of wildlife. Insects and birds that pollinate are under stress worldwide, and providing food and habitat for them allows their numbers to increase. Having a healthy pollinator population will also increase your plants' fruit production.
- Learn to tolerate some wildlife damage to your garden. There is often a trade-off between having wildlife and having a pristine garden. When the damage gets above the thresholds you can tolerate, reduce wildlife populations by limiting food and shelter.
- Never relocate wildlife to your garden. If it is suitable habitat they will eventually find it, and you will be rewarded with discovering them there. Leave animals and their families in their natural habitat where you find them.
- Never trap and release wildlife from your garden into new areas. Use licensed operators if you want to remove pest animals. Ensure that they follow the law by checking methods with animal control officers.

CHAPTER SIX

Preventing and Managing Pests

Now 'tis the spring, and weeds are shallow-rooted;
Suffer them now, and they'll o'ergrow the garden.
—William Shakespeare, *Henry VI, Part 2*

In 1830 it took four people working in agriculture to support five people outside farms—an almost one-to-one ratio. By 1930 only one person was needed to support ten others, and by 1965 one person could support forty. Now one farmer supports seventy to ninety other people. What happened? First, farmers no longer needed to rely on human and mule power. With the invention and spread of the internal combustion engine in the mid- to late 1800s, tractors, combines, and other motorized farming equipment became commonplace, making farming more efficient. Then, beginning in the 1960s, the Green Revolution sought to keep food production on a pace with burgeoning world populations through careful crop breeding and selection, improved irrigation, and increased use of synthesized pesticides and fertilizers.

While pesticides of various types have probably been used since the beginning of agriculture, postwar chemistry brought forth a number of compounds of outstanding strength and duration. In 1939 the Swiss chemist Paul Müller discovered the insecticidal properties of dichlorodiphenyltrichloroethane, or DDT, for which he won the Nobel Prize in Physiology or Medicine in 1948. Its use expanded quickly—the World Health Organization estimates that the United States produced 4,366 tons in 1944 and 35,771 tons in 1959. As demand grew, hundreds of new pesticides joined DDT in the fields as chemical companies began researching new ways to kill pests using other synthesized chemicals. But although these pesticides were usually an improvement over earlier ones made with arsenic (which disrupts several metabolic processes in animals) and cyanide (which prevents cells from using oxygen), they still had problems. Many depended on fossil fuels for production. Many killed beneficial insects as well as harmful ones. And many were directly toxic not only to insects but also to nontarget animals, such as birds. Some are stored in the fat of animals, meaning that the level of pesticides in the body becomes higher with more exposure, until it is toxic. But the new and "improved" products increased crop yields, so farmers turned away from traditional ways to manage pests, such as crop rotation. Gardeners moved right along with them.

Then, in 1962, the environmental scientist Rachel Carson published the book *Silent Spring,* warning in clear and direct language that we were poisoning the earth and its creatures with these pesticides, especially DDT. She detailed how birds were dying and waters being polluted. She discussed how even small amounts of DDT and other organochlorines, a dangerous class of chemicals that do not degrade quickly, can accumulate in tissue, how the birds that eat fish from polluted waters end up with extremely high amounts of it in their bodies, and how those high levels of pesticides lead birds to produce eggs with thin shells and high mortality. In 1972, DDT was banned in the United States, and it is now banned for agricultural use worldwide, although it is still used in many countries to kill mosquitoes that might carry malaria. A number of other postwar pesticides are also now illegal in the United States and around the world.

Thanks to Carson and other scientists, we know how harmful many pesticides can be to humans and other animals. Toxicity to humans is broken into two categories. The first, acute toxicity, is probably what alarms people the most because it means having a rapid response to exposure. However, chronic toxicity, or a response that builds up from recurrent exposures, can be worse. It may be harder to detect and may result from repeated pesticide application or rolling around on grass sprayed with toxins over several weeks. Just because you do not have an immediate reaction does not mean you have not been affected.

In reaction to these dangers, there is a huge interest in organic farming, and many people prefer to buy organic foods when possible. Organic farms seek to use earlier methods of farm management, including ecological principles and processes that support the health of soils and ecosystems, while also employing modern innovations. Organic farmers use a number of strategies to prevent pests, reducing the need for pesticides. Because this is a growing market, some companies claim their goods are organic without following organic practices, so the U.S. government has adopted strict guidelines for what types of farming practices can be so labeled, including how and which pesticides are used.

Now many people are wondering if they can apply these techniques to their ornamental plants and kitchen gardens. If farmers can live without them, do we really need harmful chemicals to manage pests in our home landscapes? How might we use ecological principles to prevent and control pests?

Integrated Pest Management

Slugs! Aphids! Black spot! Farmers and gardeners face a nearly endless list of potential foes. As pesticide use increased following World War II the emphasis was not on monitoring for and then treating pests but on routinely spraying certain pesticides at certain times of the year with the expectation that the pest would be there.

In the 1950s, some agricultural scientists began developing a method of pest control based on monitoring, timing, and targeted intervention, which they called Integrated Pest Management, or IPM. In the 1970s, as Rachel Carson's message started to penetrate our collective consciousness, IPM became the standard for many farm and landscape managers. It also became public policy in the United States when then-president Richard Nixon issued an executive order in 1972 that directed federal agencies to implement the principles. In 1979 Jimmy Carter appointed an interagency coordinating committee to oversee implementation. IPM is a huge step forward from the indiscriminate spraying of pesticides, but it is not organic—it does not forbid the use of pesticides allowed under the strict U.S. guidelines but considers them a last resort, instead preferring the least toxic treatment.

The core steps to an IPM plan, as determined by the Environmental Protection Agency, are:

- setting thresholds of allowable damage
- monitoring for pests and ensuring correct identification
- preventing pests by understanding what triggers their outbreaks
- controlling the detected pest by the least harmful methods before resorting to chemicals

Setting Thresholds

Gardeners can easily implement IPM principles. First, set the threshold you can tolerate for each insect or weed. For instance, an insect or two is not always a reason to start spraying. How much damage are you willing to bear? Are a few weevil notches in a leaf acceptable? You may be less tolerant of damage to favorite plants, and your thresholds may vary depending on where the plants are located. You might tolerate more damage on a plant at the back of the garden or in a rarely traveled side yard than on one by your front door or back patio. Many gardeners, especially those with a lawn, consider the application of "weed and feed," a mix of fertilizer and herbicide, to be a spring ritual. But think about living with a few weeds: as long as they are green, they might blend in fine—and you won't have to wonder what pesticides you and your children are absorbing when you use your lawn.

Monitoring

Once you have decided what you will accept, monitor carefully to see whether your garden's pests are increasing toward a point you cannot tolerate. Master Gardeners and extension agents can help identify which pests you have so you can treat them appropriately. You might want to keep a notebook to track the plants affected, insects found, and timing of damage. Many insects are more susceptible to control at certain stages of their life cycle, so timing and correct identification are critical to safe, effective control.

If you have one plant that is targeted first by insects or other pests, you can use it as an indicator to tell you when to start implementing controls. Monitor this plant frequently. A number of products can also help with monitoring—your garden center will have both general and specific insect traps that will help detect species and population levels.

Prevention

Just as with health care, it makes sense in the garden to prevent problems that might otherwise require drastic action. For IPM, prevention includes physical and cultural practices.

Physical barriers, such as mulch, can deter weed establishment, and edging materials can prevent lawns from encroaching on garden beds. Bands, especially with sticky substances, placed around stems and trunks may stop insects from crawling up from the ground to munch on leaves.

Good sanitation is critical to reducing insects and diseases. Remove infected leaves, even those that have fallen to the ground, and do not add them to the compost pile. Diseases can be spread with tools and equipment, so a quick wipe of your pruner's blades with alcohol or bleach is worth the time if they have been used on infected plants.

Try using cultural techniques to prevent pest infestation before implementing control. Many pests are a symptom of a stressed plant. If it does not have the right soil, water, light, or other resources, it will be more vulnerable to pests, and the pests know it. Keeping your soil healthy (see the chapter on soil) is one of the best preventive measures. Better nutrition will mean the plant is able to resist pests, and healthy soil is rich in microbes that fight plant pathogens. Avoid too much fertilizer, however, because many insects are attracted to rapidly growing shoots. Spacing plants for good air circulation will also help prevent many fungal problems.

You may also need to face the fact that you simply cannot provide good growing conditions for all the plants you want. If you consistently have a problem with insects or fungus on a plant, you will be better off moving or replacing it. Your garden may lack the light, climate, or air circulation to grow hybrid tea roses without black spot on the leaves, for instance. Maybe you could do without roses, or use shrub roses that are not susceptible to black spot. Finding the right plant for the right place is a big part of IPM.

Choose plants that are resistant to pests in your area or that help build resistance in your garden. For instance, African and French marigolds are famous among vegetable gardeners for their root exudates that repel nearby nematodes, tiny roundworms that can be serious root pests, from adjacent plants. Other plants are themselves resistant or at least unattractive to pests. If there are aphids on your roses, you can plant other species that are less attractive to them. Your garden center can help you find the right plant for the right spot in your garden.

Finally, finding the right place for the right plant is critical in preventing pests and cannot be emphasized enough. For instance, mites like hot, dry conditions. Put a susceptible species like Japanese skimmia *(Skimmia japonica)* in a sunny location, and mites are nearly guaranteed. Put it in the shade, and you will have a lovely, pest-free, low-growing shrub. Select a sun-loving plant, such as rosemary *(Rosmarinus officinalis),* or a conifer shrub for that warm spot.

Controlling Pests

Even after monitoring and prevention, you may still need to control a pest. IPM starts with the least risky techniques, which are mechanical methods to remove the pest. Some types of biological control, such as the use of ladybugs to control aphids, are available to the homeowner, but because biological control methods may carry risks, such as invasion by a nonnative species, they are often available only to professionals. If mechanical or biological control doesn't work, chemical control with pesticides is the next step. Start with the least toxic pesticides (discussed below) and move to more toxic ones if they are ineffective.

Mechanical Control Do you really need an herbicide, or can you pull a weed? These days there are improved devices for weeding, including some that you can use without stooping over. I use a "hori-hori," a narrow serrated steel blade that easily enters the soil and loosens the root so the weed can be plucked out. Could you also pick beetles off the plant rather than spray? Or shake the branch inside a bag to knock the insects off so they can be discarded? Maybe you can catch flying insects with yellow sticky traps (found at garden centers). Try these techniques first, before using pesticides.

Biological Control Some pests can be controlled by introducing another organism, usually an insect, that is either known to be a specific predator of the pest or is a generalist that eats a wide variety of species. Typically, biologists go to the home range of the pest to find its predators there, then extensively test them to see what else they eat. If tests show they eat very little other than the pest, they may be introduced elsewhere (see chapter 3 for more about this). These specialist predators are rarely available to homeowners.

Generalist predators, like ladybugs and praying mantises, are sometimes available to gardeners, but their introduction outside their native ranges is discouraged. To understand why, consider ladybugs, also known as lady beetles. Despite their dainty appearance and feminine name, ladybugs are voracious predators of aphids and other insects. Many countries, including the United States, have native species, but the Asian ladybug *(Harmonia axyridis)* is the one most commonly sold at garden centers for home release. It is usually orange-red and may or may not have black spots. In Asia members of this species gather in flocks in the fall before hibernating together in rock crevices and outcrops. The urban areas where they have been introduced for insect control do not have such topological features, so they instead often invade light-colored homes, looking for a warm spot for the winter. A great glob of swarming insects, albeit individually somewhat cute, is the stuff of horror movies. Even if you do not quiver at the sight of thousands of insects invading your home, they are a health hazard, triggering serious allergy and asthma complications in many people with their droppings and with chemicals they use for defense.

Do not release Asian ladybugs in your garden. Other species are sometimes available, but do not release them if you do not know that they are native to your area. Ladybugs tend to fly away from the property where they are introduced anyway, so your money will be wasted. In general, rather than buying even native beneficial insects, making your garden a safe place by restricting all insecticide use is the best policy. The beneficial predators will come if you have other insects for them to eat and do not spray.

Chemical Control A pesticide is any chemical that is used to kill an organism—whether animal, plant, or disease. Table salt is a pesticide when used to shrivel a slug. Herbicides kill plants, insecticides kill insects, fungicides kill fungi, and so on.

Pesticides are strictly regulated in the United States and most other countries, with explicit instructions on how they can be used and limits on what suppliers can claim about ingredients and effectiveness. The Federal Insecticide, Fungicide, and Rodenticide Act (FIFRA), administered by the Environmental Protection Agency with the assistance of relevant state agencies, requires that all pesticides be registered with the EPA after studies determine their dosage, effectiveness, and impact on the environment and on humans. The information on the label is really a legal document, and the end user must follow the instructions or risk violating the law and being fined. There are many more pesticides for agriculture than for home and garden, and their agricultural use is highly regulated, requiring complete documentation, among other things. Home use is much less regulated, but the label instructions must be followed.

All pesticide manufacturers in the United States must also produce a Material Safety Data Sheet (MSDS) that gives information about each product's toxicity and instructions on storage and emergency precautions. It is a good idea to read this before using the product. The manufacturer's website should provide its data sheets, and there are several free clearinghouse websites for the data sheets of many manufacturers (see this chapter's Resources).

Commercial products that are sold as pesticides, no matter how seemingly harmless, are subject to FIFRA. This means that suppliers of less-toxic materials labeled as pesticides that have not gotten EPA approval are in violation of the law. Many of the home products recommended from gardener to gardener and discussed in this chapter, therefore, are not legally pesticides. But because they generally have not been rigorously studied and have not been approved, they should be considered as potentially dangerous as any pesticide.

Pesticides generally contain both active and inert ingredients. The active ingredients kill the pests, while the inert ones help the active ingredients stick to their target, give the formula color, stabilize it, and so forth. FIFRA requires that labels list the names of all active ingredients and their percentages in the formula. All active ingredients must be registered with the EPA. The label must list the total percentage of inert ingredients but not their names or relative contributions. This is dangerous, because surveys have shown that consumers interpret *inert* to mean "harmless and not toxic," which may not be true—and we cannot know if we do not know the ingredients. *Other ingredients* is now the recommended term, but it is still legal to use *inert*. When reading the label, check both carefully. Most companies follow the laws, but some try to slide between them. Bill Schneider of the Biopesticides and Pollution Prevention Division of the EPA notes that some companies list compounds known to be relatively safe as active ingredients and include the true active ingredient as an inert in order to avoid having to register the true active as a pesticide. This is especially the case with many of the biopesticides, discussed below. It is confusing and complicated enough to make one try hard to prevent pest problems rather than to control them!

Because there is so much distrust of synthetic pesticides, many people are turning to the new biopesticides, derived from plant material or microorganisms. There is a good rationale behind these products. Plants naturally produce chemicals that repel insects. When these are isolated and stabilized, they can be used to control pests. People think biopesticides are safe for humans because they are from "natural sources," forgetting that there are some pretty poisonous plants out there. Nicotine, for instance, is a naturally occurring and effective insecticide—that kills humans too. The safer biopesticides include terpenoids, naturally occurring plant compounds that often have medicinal qualities, like pyrethrins, which are derived from plants related to chrysanthemums. Essential oils (concentrated liquids derived from aromatic plant compounds) from cedar, rosemary, and garlic are also effective terpenoids but work best in an enclosed space like a greenhouse. Other oils, such as cottonseed and soybean, can suffocate insects when sprayed on them.

Before a pesticide can be registered, its LD_{50} must be determined. This is the single dose that will be lethal to 50 percent of the test animals, which are usually rats or mice. If a chemical is very toxic, not much will be needed, and the LD_{50} will be low. It is expressed in milligrams of chemical per kilogram of animal weight. Tests are done for both oral and dermal toxicity. For oral LD_{50}, less than 50 is a high toxicity, 51–500 is moderate, and 501 or more is very low for humans. Dermal values are somewhat different, with less than 200 considered highly toxic, 201–1,000 moderately toxic, and over 1,000 relatively safe. Orally taken vitamin C has an LD_{50} of 11,900, so it is very safe; salt's LD_{50} is 3,000, so it is also safe but less so; and aspirin, in most of our medicine cabinets, has an LD_{50} of 1,200. It is interesting to compare pesticides by this metric. For instance, the oral LD_{50} of the synthetic herbicide glyphosate is 5,600, while that of the natural herbicide 5 percent acetic acid (also know as vinegar) is 3,310. But according to Timothy Miller from Washington State University, there is little evidence in the peer-reviewed literature to suggest that vinegar is an adequately effective herbicide at 5 percent concentration. In fact, there is an emerging consensus among weed control scientists that a minimum of 20 percent acetic acid solution is needed to kill weeds, so the LD_{50} for an effective dose would actually be much lower than that of glyphosate. This is not to say that one chemical is better than the other, simply that we should not assume biopesticides are nontoxic to humans.

All pesticides should be treated with respect. There is no such thing as a nonlethal pesticide—that is an oxymoron. We may not consider insects to be important members of the biological communities in our gardens, but the birds that eat them and that we love to watch would disagree. Many of the pesticides that affect insects also affect fish: as discussed in the water chapter, pesticides have been found in lakes, streams, and rivers, where they can make fish less able to sense predators or outright kill them, especially in combination with other pesticides, as they usually are. The bottom line is that if you are using a substance designed to kill something, you should be careful.

Anything that can be used to kill pests should be stored in a ventilated space secure from children and animals. It should be left in its original container, with the original label. When you use a pesticide, follow the dose instructions carefully and wear protective gear. Read the warning section on the label and/or the MSDS and know how to treat yourself if the chemical gets on you. Mixing chemicals is probably the most dangerous part of using them, so be very careful if you do this. Never use your kitchen utensils for measuring.

Do not spray pesticides on a breezy day—they can blow onto you or plants you do not want sprayed or end up in streams and lakes. Following application, you may need to wait for some time before touching the plants—check to see what the label says. Do not use old pesticides, and do not wash them down the drain or pour them out on the ground. Contact your local environmental protection office for information on how to dispose of them. You can find the closest hazardous waste disposal site by calling 1-800-CLEANUP or visiting www.earth911.com.

Despite all these warnings, it is important to remember that many of the harmful chemicals Rachel Carson wrote about in *Silent Spring* are no longer available commercially, though some still can be found in garden sheds and should never be used. Synthesized garden pesticides are generally much safer today, in some cases more than natural formulas such as 20 percent vinegar formulations. Still, most home gardeners do not need to use commercial pesticides or even home remedies. Practicing IPM, including manually removing weeds and insects, is possible on a residential scale, obviating the need to resort to any form of chemical warfare. Below I discuss major types of pests, with prevention methods and some pesticides that may be less harmful to nontarget organisms, starting with the least harmful (summarized in table 3). Larger animal pests are covered in the wildlife chapter.

Insects

Healthy plants are generally not susceptible to serious insect infestations, so steps to improve growing conditions are probably the most important.

Good soil and the right amount of light and water will help the most. If the chapter on water did not convince you to fertilize only when necessary, consider that insects will attack rapidly growing stems, one possible result of too much fertilizer. If you consistently have an infestation problem with a plant, carefully note its growing conditions and check with a garden center or Master Gardener to ensure it is in the best place. You might want to bring a small bit, sealed in a plastic bag, with you so that you can be sure of the species of plant and pest.

The key to effectively controlling almost any pest is to catch the infestation at an early stage. This will allow you to begin control on a small, manageable

TABLE 3. LOW-TOXICITY PESTICIDES AND HOME REMEDIES

Active Ingredient	What Is It?
INSECT CONTROL	
Insecticidal soap	Potassium salts of fatty acids derived from plant material
Horticultural oils (dormant and summer)	Petroleum- and plant-based products
Pyrethrum	Natural insecticide derived from species of *Chysanthemum*
***Bacillus thuringiensis* (Bt)**	Naturally occurring bacterium that produces chemicals toxic to some insects
Kaolin clay	Naturally occurring mineral used in toothpaste and other products
SLUG AND SNAIL CONTROL	
Sodium chloride	Table salt

infestation so you can use the least toxic methods. There are several ways to monitor, especially for insects. You can purchase traps with bait or pheromones (hormones or attractant chemicals) for some insects; place sticky traps, usually yellow because many insects are attracted to that color, around indicator plants for early detection; and construct pitfall traps—water-filled plastic cups or cans whose mouth is flush with the ground—for crawling insects. Even shaking a branch over a sheet of paper can reveal insects you might not have noticed. If you observe insects, proceed to the next stage, correctly identifying them, so you can determine if action is needed. Monitoring for weeds is easier—they just sit on the ground, waiting for you to notice them.

Uses	Cautions
Contact poison for soft insects like aphids	Can be toxic to some plants
Used to poison or suffocate aphids, thin-shelled scale insects, mites, and some caterpillars. Neem oil is effective on aphids, Japanese beetles, cabbage worms, nematodes, and perhaps some fungi. Dormant oils are applied in the winter, summer oils during the growing season.	Primarily for use on woody plants; some newer oils may also be used on herbaceous plants. May kill beneficial insects.
Paralyzes nerves of nearly all insects	One of the oldest pesticides, but it is generally safe
Especially useful for moths and caterpillars. Different strains are needed for specific pests.	Not toxic to mammals or most beneficial insects
Sprayed on plants, especially fruit trees, to repel apple maggots and other orchard pests	White residue may be unattractive
Dehydrates slugs it is sprinkled on	Could increase the salinity of the soil if used frequently

(continued)

TABLE 3. LOW-TOXICITY PESTICIDES AND HOME REMEDIES *(continued)*

Active Ingredient	What Is It?
SLUG AND SNAIL CONTROL *(continued)*	
Beer or yeast mixed with water	Homemade baits
Crushed eggshells, coffee grounds, dolomitic lime	Household waste or garden product
Copper	Store-bought or homemade bendable strips
Iron phosphate	Naturally occurring soil mineral
FUNGUS CONTROL	
Sodium and potassium bicarbonate	Sprays of naturally occurring salts related to baking soda and used in antacids
Essential oils	Concentrated liquids derived from aromatic plants such as cedar and clove
Copper sulfate	Naturally occurring salt
Sulfur	Acidic dusts and sprays long used on fruit crops
WEED CONTROL	
Glyphosate	Synthetic pesticide sold under several trade names
Concentrated vinegar	Twenty percent acetic acid solution
Corn gluten meal	By-product of corn milling

Uses	Cautions
Placed in bowls to attract and drown slugs and snails	Nontoxic to humans in moderation
Scrapes slugs' and snails' soft skin	May increase or decrease soil pH, depending on the product
Reacts electrically with mucus and prevents entry to the enclosed area	Expensive for a large area
Combined with wheat gluten to make slug and snail bait	Not toxic to most organisms
Good against powdery mildew and rose black spot	Both appear to be safe
Used for powdery mildew	May work better in enclosed places than outside
Controls fungal problems including mildew and black spot. May be used alone or mixed with lime to make the classic Bordeaux mixture used on grapes.	May harm beneficial soil organisms and some plants. Can be toxic to fish.
Controls fungal problems like mildew and black spot	May be toxic to some plants
Kills a wide array of plant species	Not toxic to mammals at normal doses. Some formulations contain compounds, especially surfactants, that may harm aquatic and other life. Does not leach into groundwater
Dissolves cell membranes. Most effective against shallow-rooted weeds.	Twenty percent acetic acid solution is very strong (compared to culinary vinegar, which is a 5 percent solution) and can cause skin and eye damage. It can also harm nearby desirable plants.
Applied before seed germination to dehydrate weed seeds	Increases dehydration in dry soil. If used on existing plants, it may act like a fertilizer. May have a strong smell and attract slugs.

Insecticides

Insecticides are not selective for harmful insects, although some might be specific to certain types of insects, such as mites. Because they also kill beneficial insects—such as pollinators and the ones that eat pests—they may make infestation problems worse. Also, insects usually have short life cycles, so there is often rapid selection for the genotypes that resist insecticides, meaning that applying pesticides may not eradicate the infestation, but will add pollutants to the area while wasting your money. If you really want to use them, though, pick the safer ones.

Systemic pesticides, such as those used in lawn "weed and feed" formulas that are dissolved in water or insecticides that are applied as soil drenches, are usually not broken down by environmental factors like rain. They can remain in the plant for several weeks, which is good for deterring insects but bad if you plan to eat the plant—or part of it—soon. Nonsystemic pesticides—like those that are sprayed on—coat the plant and kill insects on contact. Some include inert ingredients such as spreaders and wetting agents, which help them diffuse, and stickers or surfactants, which help them stay on. These additives can be at least as toxic to nontarget animals as the active ingredient. If you are going to use a commercial formula, read the label carefully. Some insecticides may have different additives and therefore also differ in toxicity and recommended use.

Insecticidal Soaps

Soaps can be effective contact poisons for soft-bodied insects such as aphids and mites. They disrupt cell membranes in the insect and perhaps also cause dehydration by removing protective wax from the epidermis. You can purchase a commercial formula or make your own 2 percent solution by mixing two tablespoons of liquid soap (not detergent!), such as a pure Castile soap, into one quart of water. Always test any solution on a small part of the plant, since soap may lead to plant injury or death. Also note that repeat applications can lead to chronic toxicity in plants, so use sparingly and with caution.

Horticultural Oils

These oils, derived from vegetable or petroleum sources, coat and suffocate insects or their egg masses. They work best on aphids, scale insects that do not have thick shells, and mites. They may also be effective on some caterpillars without dense hairs. Some will also kill beneficial insects, and they may cause plants with waxy cuticles, such as cultivars with bluish leaves, to lose their color by dissolving the wax. Like most pesticides, they should not be used on wet foliage or when rain is expected. It is important to know the life history of the insect to choose the right oil and apply it at the right time. Dormant oils are often petroleum based and are often applied to the trunk and bark of deciduous trees where overwintering insects may be hiding. Summer oils, which are usually lighter, are often derived from plants, including cottonseeds and soybeans, and may be sprayed on foliage with limited toxic effects on the plants. The popular neem oil, extracted from the fruit and seeds of the neem tree *(Azadirachta indica),* is an example of a summer oil and is effective on aphids, Japanese beetles, cabbage worms, nematodes, and perhaps some fungi. It must be diluted and does not emulsify well in water, so a surfactant is needed. It is easiest to buy it already formulated.

Pyrethrum Insecticides

Derived from several species of *Chrysanthemum,* pyrethrum disrupts the nervous system of nearly all insects, including beneficials. It is one of the oldest biopesticides and is generally considered safe for mammals but should not be used near water because it is toxic to some fish, including trout. There are synthetic pyrethroids available too. Both natural and synthetic pyrethrums should be used with caution.

Bacillus thuringiensis (Bt) Sprays

Bt is a naturally occurring bacterium that produces chemicals that are toxic to some insects. Different strains are needed for specific pests. It is especially useful for moths, including gypsy moths. Bt may affect some beneficial insects that are closely related to the target pest, but it is not toxic to other animals.

Kaolin Clay

This whitish clay occurs naturally and has low mammalian toxicity—it is even used in toothpaste and some other products. It is useful, especially with fruit trees, for repelling several types of insects, such as beetles and borers. The downside is that it gives a whitish cast to the plant, although this washes off. Kaolin clay can be sprayed onto the trunk or leaves, and it agitates insects by attaching tiny particles to them and forming a barrier to their eating or laying eggs on the plant. It is not truly a pesticide because it does not kill insects, and it can only be used as a preventive—it is not effective against existing populations.

Slugs and Snails

Slugs and snails lay their eggs, which look like gelatinous globs of pale caviar, in protected places like woodpiles, old garden pots, and mounds of leaves or mulch. Immediately remove dead leaves from beds to eliminate this potential nesting site, but if you have a problem in your mulch, it is best to replace it in the fall, after the first freeze, to eliminate overwintering eggs. Slugs and snails may be picked up for disposal, which is easy and safe. When I worked in landscaping, I had a co-worker who cut slugs in half with her pruners, which is a bit disgusting. My method, used in my former garden, which had a moderate problem with snails, was to pick them up and place them in the middle of the road, giving them a sporting chance to survive. Of course, the hot road and speeding cars meant the ones that lived would be faster and tougher. I was told this was a species of snail coveted as escargot, and a former professor of mine touted the deliciousness of slugs, but I don't plan to find out how either of these taste.

Even though I do not want to eat them, mollusks have many predators. The main reason my current garden has few gastropods is the large population of birds, which eagerly forage along the ground for these tasty treats. Ducks, geese, and even chickens are fond of snails and slugs, as are snakes and salamanders. Other predators include spiders and ants, which eat their eggs.

Molluscicides

Home remedies work well for most mollusks. If you pour table salt on them, they will shrivel. Beer, or yeast and water, in a bowl whose mouth is flush with the ground, will indeed attract and drown them. The Colorado State University entomologist Whitney Cranshaw tested many beers and found that cheaper ones, measured in what he calls Bud units, are the most effective. Do not forget to change the liquid every three days. Many people swear that sprinkling crushed eggshells, coffee grounds, or dolomitic lime around susceptible plants saves them by scraping the soft skin of the mollusks, but others claim that is false. All three items can change the pH of the soil, though, so only use them in moderation. If you have a small area or a large budget, you might place copper—either ready-made commercial products or three-inch-wide strips cut from a sheet from a hardware store—around individual plants or garden beds to repel slugs by causing an electric charge as the metal reacts with their mucus. When you place the strips, make sure that all slugs are removed so you do not corral them in with your prized plants, and do not leave gaps that they can sneak through later.

Iron Phosphate Bait

Iron phosphate is a naturally occurring soil mineral usually combined with wheat gluten to form pellets used to attract and poison slugs and snails. Such baits are a vast improvement on the older metaldehyde-based chemicals, still used today, which can poison pets, children, and wildlife and should be avoided. Iron phosphate is not toxic to most other organisms, and it is available in several commercial formulations—check the ingredients list on the package to see if it is included.

Fungi

If you have mushrooms, you might consider yourself lucky. They are the fruiting bodies of the underground part of the fungus, known as the mycelium. The mycelium may well be mycorrhizal (see the soil chapter) and helping with the growth of your plants. Check with local experts to identify the mushrooms and dispose of them only if they are poisonous and you are worried about children or pets consuming them. You can remove mushrooms by picking them, which will not harm the mycorrhizal parts. Wash your hands after handling them.

There are several microscopic fungi, however, that can cause problems. Powdery mildew is a whitish coating on the upper surface of leaves, but although unsightly, it rarely causes serious harm. Downy mildew, on the other hand, is found on the underside of leaves (think of looking on the "down" side of the leaf for downy mildew) and can kill your plants. Fungi of several different types can cause gray or tan dead spots on leaves. Black spot, the bane of rose growers, causes leaves to drop and can infect the cane as well. Rusts (and bacteria) form reddish-brown leaf spots and are more serious and more difficult to control.

Many fungal problems can be prevented. Avoid watering late in the day when leaves may not dry out before nightfall, or use an irrigation system that does not get leaves wet. Space plants to allow good air circulation. Remove leaf litter that may contain spores, and spread mulch in the fall to keep rain from splashing spores in the soil onto your plants. If cultural practices fail to correct a problem, replacing the plant with fungal-resistant varieties will be the best option.

Roses, especially hybrid tea roses, are prone to fungal black spot and mildew. Charlie Shull, a resourceful employee of a nursery north of Seattle, swears that one reason is that roses in mild climates never go fully dormant, so they keep even diseased leaves all winter. His remedy: first, reduce growth by using no fertilizer after late summer. Discourage flowering in the fall by not deadheading; then, as they go dormant, remove all leaves from the plant and the ground. In the fall, use fresh mulch to keep whatever spores are on the ground from getting onto the plant. Apply potassium bicarbonate as described below during the growing season if problems arise. There are, however, some great roses that resist black spot and mildew—ask at your garden center.

Sodium and Potassium Bicarbonate Sprays

Sodium bicarbonate is better known as baking soda, and potassium bicarbonate is a similar salt. They affect fungal cell walls but appear to be safe for animals. Potassium bicarbonate is particularly effective on powdery and downy mildews. Both fungicides have been found to combat black spot on roses. A bit of soap can help a bicarbonate stick to the leaf while attacking the fungus itself. Add four teaspoons or a heaping tablespoon of bicarbonate and one-half teaspoon of liquid soap to one gallon of water and spray the solution on your plants about once a week. Alternatively, you can use a commercial formula.

Essential Oils

Neem oil (discussed above as an insecticide) has been shown to prevent and control powdery mildew. Clove *(Syzygium aromaticum* or *Eugenia aromatica)* and cinnamon *(Cinnamomum zeylanicum)* oils have also been investigated for fungicidal and insecticidal properties, but whether they are effective where air circulation is greater than in a greenhouse is still questionable. Recently, an emulsion of water and knotweed (a very weedy species of *Fallopia,* some-

Tea Time?

Recently there has been much interest in compost "tea," a simple mixture of compost steeped for several days in water. It may be aerated to form compost "tea" or allowed to ferment anaerobically (without oxygen) to become compost "extract." It is then watered into the soil or sprayed onto the leaves. Aeration simply accelerates the fermentation process, so the tea is just like normal compost, as is the anaerobic extract watered onto the soil, and both take more work. There are many claims about its ability to fight diseases when sprayed on leaves, but scientific studies are mixed in their results. Part of the inconsistency results from the differences in the composition of the tea. Composts vary in salt concentration, pathogens and other microorganisms, and amount of oxygen, so the products made from them differ too. Microorganisms can also vary between a tea and the compost used to make it. In 2004 Audrey Litterick and her colleagues reviewed several studies distinguishing between aerated teas and nonaerated extracts. They noted that few peer-reviewed studies have been done on aerated tea and found that some studies using nonaerated extracts documented disease control, but results varied based on the factors above and the disease organism. The review article also mentions the transmission of human pathogens, including fecal coliform, through both teas and extracts. At this time compost tea cannot be recommended to control diseases. Use compost, but hold the tea.

times put in the genus *Polygonum*) was found to be a promising fungicide, and it may be commercially available soon.

Copper Sulfate

Copper sulfate controls several fungal problems, including mildews and leaf spots, especially on grapes and other fruit crops. When combined with lime it is called a Bordeaux mixture, which is the common form used on fruits. Depending on the fomulation, it may be applied as a spray or a dust. Copper sulfate negatively affects earthworms, which may decrease your soil health, and it may be toxic to some plant species. It can also accumulate in water and harm fish. Use commercial formulas with special care.

Sulfur

Sulfur dusts and sprays are some of the oldest fungicides in agriculture. They are used to treat powdery mildews, black spot on roses, and a number of fungal problems on peaches, cherries, grapes (although they may be toxic to the native American grape species), and other fruit crops. Sulfur is acidic and may be toxic to some plants, especially on hot, sunny days. Use a commercial formulation and apply sparingly until you know how it affects the plant. A wettable powder may be safest.

Weeds

Preventing the spread of weeds is critical to effectively managing them. Mulches, discussed in the chapter on soil, are essential, as is removing weeds before they seed. Because seeds can persist in the soil for many years, preventing plants from going to seed will save you years of work. If you don't have time to dig out the plants, at least cut off the flowers and fruits until you can do more. In the future, you will be glad you did!

Mechanical Control

You can effectively control most weeds by pulling them or digging them up. Many look at weeding as a chore. I find it to be a sort of meditation and use the time to sort out problems—my hands keep busy with a methodical task, and my mind is free to think about issues. Try this the next time you have a vexing problem, and you may be surprised! Annual species usually have weak root systems and are not difficult to pull. This is usually easier if the soil is wet, so weed just after rain or irrigation. Perennial species and woody plants may have long roots or rhizomes that are difficult to dig out, presenting more of a challenge. But sometimes just removing as much as you can will be sufficient to kill the rest of the plant: continually eliminating the parts above the ground will eventually use up the nutrients stored in the roots. I found this to be effective in controlling hedge bindweed *(Calystegia sepium),* but it requires a lot of patience.

When I first moved into my new house with its nearly three-quarter-acre garden, there were only five cultivated plants that I identified as regional wildland invaders needing removal. That was the easy part. Much of the garden, however, had been neglected for a few years, and weeds had moved in with a vengeance. Removing all of them looked to be a daunting task. Fortunately, I knew about the Bradley sisters, and they inspired my strategy.

Joan and Eileen Bradley lived in Australia in the 1960s. They owned about forty acres of bushland, which they were able to restore to its native condition with just a few hours of labor a week, using ecological principles that challenge our instincts. Most of us, when presented with a lot full of weeds, will head to the worst area first. But it makes more sense to work first in the areas that are the closest to pristine. This is especially true in wildland restoration: if you remove the few problem plants in the best areas, the remaining native species will be released from competition and can quickly recover and produce seeds. Starting in the parts of their property containing the most natives, Joan and Eileen walked their dogs in the morning and afternoon, removing weeds as they exercised their pets. As those areas recovered and sent more seeds raining down into nearby areas, promoting natural regeneration, the sisters expanded their walks bit by bit.

My garden is not a wildland, but by using the Bradley method I have slowly recovered it by myself, also with just a few hours a week. The first few years, I often grew impatient and got ahead of myself. The garden had been planted with many wonderful species that were in some cases completely overgrown. Hacking through the blackberries to discover the secret garden beneath was so exciting that I wanted to do that instead of the boring removal of germinating seedlings in previously cleared areas. This led me to modify the Bradley method slightly. While I concentrate my "weed patrols" on the cleanest areas, I also make a sweep through the still-weedy areas to remove plants that are going to seed.

Weed Disposal

Once you have removed your weeds, what do you do with them? Should they go into home compost, municipally collected green waste, or trash? It depends. (Isn't that usually the answer?) You will need to consider the species, whether or not there are seeds, and how much matter there is. Most herbaceous species are easily composted, but woody branches break down more slowly and test the patience of gardeners wanting to use their compost and are probably best given to large-scale municipal programs. It is best to remove weeds before they go to seed. Keep in mind that some species, like those in the sunflower/daisy family, may continue to form seeds even after the plants are pulled up. If your weeds have seeds, it is better to put them in the municipal collection, if available. The city's commercial composting is hot enough to kill seeds, but your home composting is not. If there are rhizomes in your weeds, do not put them in your home bin—they may start growing and ruin the compost. If you lack other options, you can put the weeds in a black plastic bag and leave it in the sun for a few days to try to kill the seeds and rhizomes. In most cases, yard waste can be recycled through composting, so it should not be included with the trash. Why add to already crowded landfills and other long-term disposal operations?

If you have a large lot with several woody plants that need removing, renting or buying a shredder/chipper might be a good idea. Arborists use huge and somewhat dangerous ones for large trees, but small, inexpensive ones work for less material and are relatively safe. As you remove the plants, you can run them through the chipper and have instant mulch that will help keep some seeds in the soil from germinating. My husband loves power tools, so that is his garden task. If there is green waste collection in your area, either curbside or at a central site, that may be the best alternative. Large-scale composting facilities are equipped to deal with woody material.

One garden I toured had a clever idea for dealing with narrow woody sticks. The owner had driven a series of large, sturdy poles into the ground in two rows several inches apart. She then dropped sticks between the rows. As they decomposed, they settled to the ground and she added more to the top. This made an attractive rustic fence that divided spaces in her garden and cleverly recycled her yard waste. Some artistic gardeners—or their creative children—weave vines and other parts of removed weeds into balls, teepees, or other transitory pieces of art. You can also use smaller branches for stakes and compost the leaves.

One method of weed control that requires little disposal utilizes a weed torch. If you have pavement or paving stones and want to kill all of the plants in the cracks and crevices, torches are a good nonchemical option. (Obviously, one needs to be careful when using fire.) There are several weed torches commercially available. The object is not so much to burn up the plant as to burst the cells. Even with fire, however, deeply rooted species may require several applications.

Herbicides

Most home gardeners do not need to use herbicides, even the "safe" alternatives discussed here, or other home remedies. If you want to use them, however, apply them in ways that minimize the amount that gets on other plants or the ground. One popular technique is the cut-stump method: the gardener removes most of the aboveground part of the plant, leaving a small stub or trunk, then carefully brushes or drips the herbicide onto just the freshly cut area. Another option is to cut a "frill" or make several shallow cuts through the bark into the living cambium, then brush or squirt herbicide onto it. These methods are not effective for all plants, but they can kill many herbaceous and woody invaders.

Glyphosate One synthetic herbicide that is effective and apparently nontoxic to mammals is glyphosate, sold under several brand names (read the label to see if it is listed as an active ingredient). Glyphosate works on plants by disrupting a pathway that produces amino acids. Animals, including humans, do not have this pathway—we get amino acids from our food. Glyphosate is a systemic pesticide and kills a wide array of plant species. In the soil it is quickly degraded by microbes. However, many formulations contain compounds, especially surfactants, that may be harmful to aquatic and other life because they affect cell membranes. Glyphosate should always be used carefully.

Vinegar Vinegar has become popular with many gardeners who believe it is harmless to humans because it is used in the kitchen. It may cause some leaf and shoot dieback, but it may not affect the roots—culinary vinegar is just 5 percent acetic acid, and as noted, consensus among weed control scientists suggests that solutions with less than 20 percent are not effective for weed control. Tim Miller of Washington State University tested 5 percent solutions of white and cider vinegar on annual weeds (which, because of their weak roots, would be the most susceptible) both early and late in the season and got no more than about 15 percent weed mortality. The 20 percent formula, created for weed control and available at some garden centers, is an extremely corrosive acid that can cause serious eye and skin damage, so all applications should be done with the same level of protection that you use for other harmful chemicals. Be aware that taprooted weeds will likely survive even 20 percent solutions and at best will require several applications. Control improves with higher air temperature for both 5 and 20 percent solutions.

Corn Gluten This is a by-product of corn milling, so it gets high marks for using a waste product. If applied before germination it will dehydrate seedlings, but because it is 10 percent nitrogen, if used after germination it will act as a fertilizer and increase the vigor of the weeds. It is most effective if the soil is dry, but Tim Miller says that he rarely sees it result in greater than 20 percent weed mortality. Corn gluten may also have a strong smell, which usually dissipates quickly. Miller also notes that it may attract slugs.

Final Thoughts

When I worked for an interior landscaping company in the early 1980s, one of my responsibilities was to keep the greenhouses and our large installations free of insect pests. I was a professional applicator, licensed by the state and knowledgeable about chemicals and their regulations. I always had respect for the insecticides I used and applied them while wearing full protective gear and in complete compliance with the laws, but in other ways I was remarkably casual. Granted, this was before the days of strict rules about reentry into sprayed areas, but we generally allowed reentry as soon as the leaves were dry. I recall driving back from another city in the company van after spilling a widely used but potent insecticide. We had all the windows open, but the noxious fumes were powerful and probably very unhealthy.

We were working on large commercial installations—a scale that makes mechanical controls very expensive—but now that more than twenty years have passed, I can see how unnecessary our reliance on chemicals was. There were a number of ways we could have handled pest control, including insecticide applications, in a manner safer to both our clients and ourselves. Even though we were aware of IPM and practiced it to some extent, we generally took a zero-tolerance approach to pests.

I have not used insecticides of any sort for many years. I maintain healthy soils for healthy plants. I work to increase the carrying capacity of birds in my garden, and they help control insects and slugs. Because I do not use insecticides, I also have predatory insects. I can honestly say that with nearly an acre of garden, I do not have any insect situations that are even close to a problem. I confess to using some herbicides, especially on hard-to-control woody plants that sprout from stumps, but I always apply the herbicide directly to the cut stump, minimizing the amount that may contaminate soil and nontarget organisms. I am reaching the point where large plants of these species are controlled and I just need to pull up seedlings.

Take a look around your garden. Are there any pests you can control without pesticides? How about all of them? Are there any plants with a consistent problem? Can you replace them, or at least move them to a more remote place where you can tolerate the pests? Ask yourself these questions, and you are on the road to IPM.

Guidelines

- Decide how much damage from each pest you can tolerate. This may vary in different parts of your garden and for different plants. Monitor the pests to see if they are increasing, noting where they are and if they increase at certain times.

- Adjust cultural conditions to increase the health of the plant; make sure it is in the right place for optimal growth. Clearly tell garden center staff about growing conditions when trying to find a plant for a specific location.
- Prevent pest problems before they occur by using such things as physical barriers to root and rhizome encroachment, fences, and repellents.
- Always practice good sanitation. Although fallen leaves are good mulch and soil replenishers, you should remove leaves that may carry diseases and debris that allows insects, slugs, and snails to hide. Keep your tools clean to prevent disease spread.
- If you choose to control a pest, try the least toxic methods before resorting to chemical warfare.
- Allow beneficial insects such as pollinators and insect predators to help with your pest problem. Use pesticides only when all other steps have failed and you cannot tolerate the pest. Strictly follow the label directions. Store pesticides in a safe place and use only when the label is readable. Avoid overapplication.
- Prevent seed production by weeds. You will be so happy you did in a few years! Sometimes repeatedly removing all the aboveground parts will weaken the plant to death.
- Dispose of weeds appropriately. Remove seeds and put them into the trash or municipal yard waste before adding leaves and stems to the compost bin or chipper/shredder. Never dump weeds into nearby woods, fields, or aquatic systems.

CHAPTER SEVEN

Confronting Climate Change

Each one of us is a cause of global warming, but each one of us can make choices to change that with the things we buy, the electricity we use, the cars we drive; we can make choices to bring our individual carbon emissions to zero. The solutions are in our hands, we just have to have the determination to make it happen.

—Al Gore, *An Inconvenient Truth*, 2006

These days it seems like you cannot pick up a newspaper without finding another article on climate change. We have been talking about it for many years: the Kyoto Protocol to establish guidelines for reducing greenhouse gas emissions, the culmination of years of scientific study and governmental negotiations, was adopted in 1997 (the United States is notably not a party to the agreement). The public discusses it, politicians debate it, but little progress is made: according to the Fourth Assessment Report (AR4) of the Intergovernmental Panel on Climate Change (IPCC), temperatures and sea levels are still rising and precipitation patterns are still changing. Aldo Leopold was also frustrated by the slow rate of change in response to environmental problems in his day. In perhaps the most eloquent essay in *A Sand County Almanac,* "The Good Oak," he traces through tree rings on a stump on his Wisconsin property the history of ecological harm and progress over the seventy or more years the oak lived. He notes the horrifying slaughter of game bird species and the extinction of the passenger pigeon but also the gradual progress of innovations such as the first Arbor Day and land stewardship education for farmers. Perhaps such incrementalism is inevitable in creating the will for change, but with climate issues we cannot delay our solutions.

Alterations in our climate have happened before in Earth's history, but this is the first time that humans are the main cause. There are many contributing factors, but the Industrial Revolution, which started in about 1750 and has depended heavily on fossil fuels such as coal and oil, is the biggest one. Factories, electricity generators, vehicles, and other staples of modern life are all implicated in releasing large amounts of greenhouse gases.

The gases that contribute to climate change include carbon dioxide, methane, nitrous oxide, and even water vapor. Carbon dioxide (CO_2) is most often mentioned, because it is increasing rapidly and remains in the atmosphere for a long time. Among other methods, it is created from the burning of fossil fuels. Plant decay also produces CO_2, but it is counterbalanced naturally through the uptake of CO_2 by growing plants, which use it for photosynthesis. However, burning coal and oil—fossilized plant and animal matter—creates larger amounts of CO_2 than terrestrial and marine plants can use.

Methane, another greenhouse gas, is produced by decaying material in landfills, wetlands, agriculture, and other sources. Cattle emit large amounts of methane during their digestive process and globally account for 28 percent of human-related methane production. The chemical destruction of methane in the stratosphere, earth's upper atmosphere, produces another greenhouse gas: water vapor.

Other contributors to climate change include nitrous oxide, created by, among other processes, fertilizer use and fossil fuel consumption, and aerosols, which are tiny particles suspended in a gas, often natural substances such as dust or very fine ash from fires or volcanoes. Ozone, a form of oxygen, is produced and destroyed naturally through chemical reactions, but hydrocarbon-based gases increase its production. Ozone is both harmful in the near atmosphere and helpful in the outer stratosphere, where it reflects energy into space. Freon and other halocarbons employed in refrigeration and industry destroy the stratospheric ozone layer; international regulations have decreased their use.

Each of these gases lasts in the atmosphere for a different length of time, from methane, which has a lifetime of about 12 years, to nitrous oxide, which persists for about 110 years. Carbon dioxide does not have a well-defined lifetime, but according to the AR4, about half of new carbon dioxide is removed within 100 years, and 20 percent may stay around for thousands of years. The AR4 authors report that only the complete elimination of human-induced emissions would allow the atmospheric carbon dioxide level to stabilize, but small to moderate emission reductions may at least slow the future rate of growth. It is going to take radical cuts in emissions to make a substantial difference.

As the debates over whether or not climate change is real and what is causing it subside, most of the discussion is shifting to what and how great its impacts will be and how we might stop it or at least lessen its effects. While some gardeners might first cheerfully contemplate all the semihardy species they could soon be growing, there are sobering consequences we must face.

According to the AR4, eleven of the twelve years from 1995 to 2006 were among the hottest on record since 1850. How great further temperature change will be depends on whether we can reduce the amount of greenhouse gases, especially carbon dioxide, in the earth's atmosphere. The IPCC report, the work of hundreds of scientists from around the world, says that the likely temperature increase now projected during the twenty-first century, if we allow carbon dioxide levels to double over current conditions, is an average of 3.6° to 8.1°F higher than temperatures in the 1980–1999 period. A most likely increase would be around 5.4°F. In addition, while scientists are confident that overall the globe is warming, higher temperatures will cause unstable weather patterns, so at times regional temperatures will fall below normal. It is also important to remember that the figures are averages, which means in some regions the changes will be more severe and in others less. These effects may be greater in the Northern Hemisphere, where there is more land, which gets hotter than the ocean.

Increasing temperature, while it often gets much of the press, is just part of the story. The IPCC scientists are highly confident that regional-scale

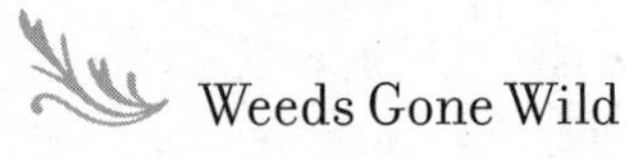

Weeds Gone Wild

Carbon dioxide is essential for plant growth, and some studies have suggested that elevated carbon levels lead to more robust invasive plants (and perhaps natives). Thomas Sasek and Boyd Strain found more growth and branching in the introduced Japanese honeysuckle *(Lonicera japonica)* than in a native coral honeysuckle *(L. sempervirens)* when both were exposed to more carbon dioxide. In an earlier experiment of theirs the aggressive invader kudzu *(Pueraria montana* var. *lobata)* increased its biomass by as much as 51 percent after sixty days. It is worth noting that in these two experiments the plants were grown alone. In diverse communities where they have to compete for resources, they may not be as vigorous.

Increased carbon dioxide will also turn certain species into greater pests. For instance, some plants will increase their pollen production, a potential problem for allergy sufferers. Poison ivy will grow more vigorously and produce stronger urushiol, the oil that causes rashes. Some plants will develop a thicker leaf cuticle, which means that herbicides will have a more difficult time entering the leaf—perhaps leading desperate gardeners to use more, and with higher levels of surfactants.

changes will also affect the amount and, possibly, timing of precipitation. Some areas, such as the western United States, the Mediterranean, South Africa, and parts of Brazil, will probably become more arid. Hundreds of millions of people will experience a lack of water. Arid conditions could lead to more frequent and intense wildfires, as dry vegetation builds up. In eastern North and South America, the moist tropics, and higher latitudes, however, there will probably be increased precipitation and storms. Heavy precipitation events will increase soil erosion and damage crops.

Coastal areas will be especially hard-hit. According to the AR4, the number of high-intensity storms, such as Category 4 and 5 hurricanes, has increased globally by 75 percent since 1970, with regional variation, and IPCC models forecast that the trend toward stronger storms will continue. Melting polar ice and glaciers will cause coastal flooding and could eliminate some low-lying islands and coastlines. Scientists now project that about 30 percent of coastal wetlands, rich habitats that are home to many animals and that help mitigate storm damage, could be lost. Irrigation water may become saline from coastal contamination, killing crops. Oceans are expected to become more acidic, which will affect coral reefs, shellfish, other shell-forming organisms, and their dependent species.

Floods, severe storms, and intense heat waves, coupled with an increase in diseases and their vectors (such as rodents and mosquitoes), will all lead to human deaths. Loss of croplands will be accompanied by a decline in the yields of cereal grains such as wheat, rice, and corn, the staples of most diets, potentially shifting growing regions and supply chains. Decreased snowpack in the mountains of North America could lead to decreased water for agriculture, drinking, and other uses in nearby cities. At the same time, rising temperatures will lead to increased water demand and potentially to water quality issues such as eutrophication from increased algal and other plant growth.

The IPCC scientists estimate that 20 to 30 percent of the plant and animal species they have so far assessed risk extinction if global temperatures increase 2.7° to 4.5°F—and as stated earlier, they think an increase of 5.4°F or greater is likely. The stories of polar bears drowning because they have to swim farther to ice floes will become more common, and other animals will also have problems if they cannot move to new areas.

Plants are moving as well. Jonathan Lenoir and his team studied 171 forest species in western Europe and found that 118, including low-elevation species, are shifting into higher, cooler locations by an average of 95 feet of elevation per decade—a huge amount for plants. Herbaceous species, with shorter generation times, are generally more able to relocate than woody plants, which may not begin to reproduce for ten or more years after seed germination. Movement in some areas is blocked by urbanization, which cuts down on sites for germination. All of this means that many plant species, especially those in montane and arctic areas with a limited ability to migrate to cooler climates, will go extinct. Hot spots of biological diversity—like California, which has 2,300 species that grow nowhere else on earth—are of particular concern. They often have a unique topography, climate, or soil type, so their plants will not be able to migrate to a similar area.

Plants are also changing their phenology—the timing of events such as spring leafing and flowering and fall leaf drop—in response to climate change. Abraham Miller-Rushing and Richard Primack compared the cur-

Project BudBurst

Around the world, there are many programs tracking plant phenology. In the United States, the Chicago Botanic Garden has spearheaded Project BudBurst, for which volunteers record the first bud break, first bloom, peak flowering, first ripened fruit, and amount of fruit for several native species. While raising awareness about climate change and plants, it also provides information about how phenology is changing across the country. It is easy to help—go to the Project BudBurst website listed in the Resources and click on Participate!

rent phenology of plants in the Boston area to records from the 1850s by Henry David Thoreau and found that plants are flowering from four to thirty days earlier in response to a 2°F temperature increase, with a mean of seven days. What will happen with a 4–6°F increase? Although earlier flowering and leafing may seem like a trivial problem, remember that plants are part of integrated communities. For instance, in northern Europe the English oak *(Quercus robur)* now leafs out two weeks earlier than previously. The leaves are a food source for the winter moth *(Operophtera brumata),* and the moth's reproductive cycle has adapted to produce larvae two weeks earlier. However, the pied flycatcher *(Ficedula hypoleuca),* a migratory bird that feeds on the moth larvae, continues to arrive at the previous regular time, when the larvae are declining. As a result, the bird population is dropping in response to this reduced food source.

What Can Gardeners Do?

With so much grim news—the potential changes and the difficulty in reducing greenhouse gases, especially carbon dioxide, in the atmosphere—it is easy to despair and maybe just hope that future generations will figure it all out. That is not an option. We need to take steps, from tiny to drastic, to fight climate change now. Gardeners are in a great position to make a difference. There are many things we can do to help plants and animals adjust to changing conditions and to reduce the magnitude of climate change.

Help Plants and Animals Adapt

In our highly fragmented landscapes, corridors that allow species to move to cooler areas are important. Parks and home gardens can serve as pathways to higher altitudes or latitudes, especially if they are large enough to accommodate migrating animals. (See the wildlife chapter for suggestions on improving the habitat value of your garden.) If a native plant species from an apparently wild source (rather than from a stand of cultivated plants on or near your property) pops up, consider adopting it in your garden as a migrant.

If plant species cannot manage to move to appropriate locations on their own, scientists may decide that their plants or seeds should be carried to safe sites. There is still much scientific and philosophical debate about this, but interested gardeners may wish to look into helping with "assisted migration" programs. These coordinated regional plans are run by universities and botanic gardens around the world. The Center for Plant Conservation is a network of such institutions across the United States. Under no circumstances should an assisted migration be done without extensive scientific evaluation beforehand—the consequences of unplanned introductions could be disastrous. Many botanic gardens also maintain seed banks for potential future research and restoration and may need help collecting seeds.

Grow More Trees

Plants use photosynthesis to convert carbon dioxide into sugars and other compounds they need. Therefore, more plants mean more carbon taken up, or sequestered, from the atmosphere. The IPCC estimates that more tree plantings, forest preservation, and improved agricultural methods could offset 10–20 percent of the world's fuel emissions. The Environmental Protection Agency calculates that forests and croplands sequester 12 percent of total CO_2 emissions and urban trees up to 2 percent more, so this seems a realistic goal. In fact, the U.S. landscape is considered a carbon sink, which means that it sequesters more carbon than it produces. But unless we act now this will stop being true, due to increased harvesting of trees, changes in soils (see below), and maturing trees.

The rate of sequestration varies by tree species and age, soil type, climate, topography, and management, so check with local garden centers and extension agents about which species are best for your region and your garden. Only growing trees accumulate carbon, and only those with large trunks hold substantial amounts. Mature trees sequester less carbon, and when they die and decay they release most of it back into the atmosphere, so keeping your landscape relatively young has the greatest benefits, although it is impractical for most people. Deciduous trees are somewhat more effective at sequestration than evergreens because they need to produce new leaves each year. Fast-growing deciduous trees may sequester 70 pounds of carbon each year, moderate growers 37 pounds, and slower growers a mere 17 pounds. Scientists are also breeding fast-growing hybrids that should store more. But here is the reality check: according to the World Resources Institute, the estimated annual carbon footprint of each American is about 23.5 *tons.* You will not be able to balance your carbon footprint simply by planting trees, but they have many other benefits.

Putting trees around your home can have a major effect in reducing your household energy use. In the United States alone, about one-sixth of all electricity generated, representing about 88 million tons of carbon per year, is used to cool buildings. As anyone who has picnicked under them knows, trees are effective at blocking solar radiation—up to 30 percent of what would otherwise reach the ground—which can decrease heat buildup inside buildings, leading to a reduction of up to 25 percent in air conditioner and other energy use and therefore lower demands on power plants. This in turn means lower carbon dioxide and nitrous oxide production, which reduces ozone levels. Trees can also reduce air temperature by 2° to 4°F in their immediate vicinity by transpiring water taken in by the roots through pores in the leaves.

Which trees are best for cooling your home? It depends on your region, and local garden centers can help you choose, but in general, deciduous species are best in the temperate zones because they provide shade in the summer and in the winter allow radiation through their branches to warm the house. However, if protection from winter winds is important, a row of evergreens would work well. To cool a building, plant trees near its warmest sides, which at northern latitudes are south, west, and, in some areas, east. Shading windows and air conditioning units also reduces energy use. Trees should be located far enough away to prevent root damage to the house but not so far that shade fails to reach the building. Also keep power lines in mind: too-tall trees or those with branches that tangle with lines can be disastrous in storms.

A green roof, discussed in the water chapter, may help keep the building cooler too. Keep curtains closed during hot times and open windows instead of running air conditioners to reduce energy use. Houses in hot climates used to be designed for maximum cooling, with stucco or strategic ventilation, but since the invention of the air conditioner we no longer seem to follow energy-efficient building principles. Consider cooling and heating when designing a new home or addition.

Leave the Soil Alone

Trees are also useful in preventing soil erosion. It turns out that soil is the greatest terrestrial carbon sink (the oceans have it trumped, though), holding 1.65 trillion tons! Soil naturally accumulates carbon—all the organisms in it are teeming with the stuff—but it loses some when it erodes or is otherwise disturbed. Studies such as those by Kern and Johnson in the United States have found that the surface carbon content of soil is significantly less after tillage. Rattan Lal has linked good soil practices to the sequestering of 5–15 percent of annual fossil fuel emissions. Do not disturb soil more than is necessary for planting. Most people do not need to use a rototiller, even before planting a food garden. Soils that demand that level of tilling are likely not suitable for food gardening anyway. Lawns may be tilled to remove large areas of turf.

Use as little fertilizer as possible—none is best. Fertilizer is one of the main sources of nitrous oxide, naturally produced through breakdown of nitrogen by soil microorganisms. Although organic fertilizers are more climate-friendly than synthetic fertilizers, which require more energy to produce, both produce nitrous oxide. See the soil and water chapters for other reasons to use fertilizers conservatively.

Use Your Muscles

There are millions of two-stroke engines in use in lawns and gardens in the United States. These noisy engines are found in lawn mowers, chain saws, leaf blowers, string trimmers, rototillers, and other types of equipment. (You can also find them outside the garden, in scooters, some water vehicles,

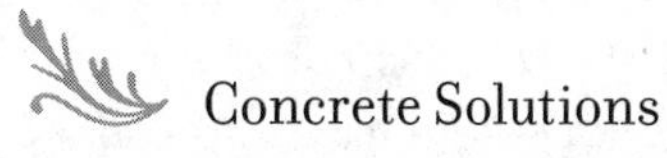

Concrete Solutions

Gardens contain much more than plants, and most gardeners must invest in patios, walkways, retaining walls, or other types of hardscape. One of the least expensive—and therefore most ubiquitous—options is concrete. Perhaps because it is the most commonly used building material, concrete is responsible for 7–10 percent of CO_2 emissions worldwide, the most after transportation and electricity generation. While quarrying for such materials as sand and gravel is responsible for some emissions, the biggest culprit is portland cement, a basic ingredient of concrete, mortar, and stucco. Portland cement was patented in 1824 in England and named after an island where a similar natural stone is found. It is created in kilns by heating limestone and clay to temperatures of 2,500–3,000°F. You can imagine the energy required for that! The fossil fuels used in making portland cement release a considerable amount of carbon dioxide. There are less-energy-intensive alternatives, such as magnesium or aluminum cement, and many of these concretes are at least as hard as those made from portland cement. If you are planning to use concrete in a project, discuss your options with local green builders.

and model planes.) Two-stroke engines were invented around 1880 and are so called because they complete the fuel intake, compression, combustion, and exhaust cycle in two strokes of the piston instead of the four of other engines. They are lightweight and relatively powerful, but one two-stroke engine can produce more emissions in a given period than thirty to fifty four-stroke engines, such as those in cars. The Environmental Protection Agency developed rules mandating a 70 percent reduction in emissions of hydrocarbons and nitrous oxide in new engines by 2007, but most equipment in use is older than that. And even if emissions are reduced, the gas itself can be a pollutant: people often spill when filling the fuel tanks, and this gas can end up in groundwater.

Do we really need to use motors to garden? The answer is, in most cases, no. How about a broom instead of a blower to clear driveways and patios? (Just remember not to clean them with water—see the water chapter.) A rake and a good old-fashioned push mower for the lawn? Designs have improved so mowers are easier to use, and they provide good exercise. Plus, they leave clippings behind, which decay and feed the grass.

However, as you may recall from other chapters, lawns (not including native grasslands and meadows) are not good for many conservation reasons—fertilizer and pesticide use and runoff into aquatic systems, high water consumption, lack of habitat or food for wildlife, and so on. Minimizing a lawn's footprint as much as possible will have many positive effects—including on climate change. Yaling Qian and Robert Follett found that turfgrass sequesters only about 800 pounds of carbon per acre per year. According to the North East State Foresters Association, forests in their region have approximately 238,000 pounds (119 tons) of carbon per acre in living and dead trees and soil. Now, you are not going to convert your lawn to a forest, but the point is that trees and woody plants are better at holding carbon. If you get rid of your lawn you will improve the environment, and you will not need a mower.

If you absolutely must have power tools, try switching to electric. While about 70 percent of electric power in the United States comes from burning fossil fuels, especially coal, the generating plants that do this are usually more efficient than the engines in gas-powered garden equipment. Learn how electricity is generated in your area. If it is from fossil fuel, be especially diligent about reducing electrical use. Do what you can to support renewable energy for your community.

Be a Locavore

In 2007 *locavore* was the *New Oxford American Dictionary*'s word of the year. It has since become a movement. The word is a combination of *local* and *-vore,* from the Latin *vorus,* which means "one who devours," as in *omnivore* or *herbivore.* The idea is to eat mainly food grown or produced within one hundred miles, to reduce the amount of fossil fuel burned to get it to market. Rich Pirog and colleagues found that most fresh food travels an average of fifteen hundred miles to get to our plates, a stunning figure. Pirog and his team have also shown that the conventional food supply system releases four to seventeen times more CO_2 than eating within the one-hundred-mile radius. There is variation because trucks produce more CO_2 per food unit than container ships or rail cars, but the message is clear: buying locally or growing your own food releases less CO_2. Eating locally keeps us in tune with the seasons too, since we will tend to eat what grows in our region during that period. Because it is fresher, the food may also be healthier, since nutritional content can decline over time. It may even be better for your health and the environment to buy local nonorganic food instead of organic food shipped from far away. If your store does not have much local produce, try farmers' markets and fruit and veggie stands or join a local community-supported agriculture (CSA) program that works with farmers to deliver fresh, seasonal produce to your home.

You may also enjoy growing fruits and vegetables in your own garden, the ultimate locavore experience. There are numerous books on the subject (a few are listed in the Resources), and there are even people who will design and maintain your veggie beds if you don't want to do the work. Some communities have programs that encourage the use of otherwise vacant land for food gardens. The Green-Up program in the Bronx, run by the New York Botanical Garden, uses lots between houses. Other places, such as Detroit, let unemployed workers grow vegetables on public land, and a program in Seattle matches people who want to grow vegetables but have no land with those who have land but are unable to garden.

It is best to concentrate on growing what will do well in your area rather than setting your mind on particular varieties. Local extension agents and Master Gardeners can help you choose. What your family likes to eat and you enjoy growing is important too. My husband, Brian, is half Irish and half German, so potatoes are in his heritage. I love digging for them (buried treasure) and he likes eating them, so potatoes take up much of our vegetable plot. I also grow many fresh herbs for cooking.

Food plants are usually fast-growing species, producing lots of leaves, roots, or fruit. But rather than falling to the ground and replenishing the soil, those materials are removed for *our* nutrition. All that fast growth requires not only extra fertilizer but frequent irrigation. While I generally discourage these, a food garden is one place where they are unavoidable. The benefits of eating so locally that you need just walk out your door for food outweigh the negatives. Use compost and well-rotted manure to feed the garden and water efficiently—in the morning, watering only the garden bed—as needed. The few leaves of garden trees that I rake up also go onto the vegetable plot in the fall, to be spaded in before spring planting.

While many grocery stores are responding to the locavore movement by labeling produce with its origin, garden centers have been slower to catch on to the desire to buy local. Most do not have the facilities to grow their own ornamental plants and must rely on wholesale growers that are often located in areas with good outdoor growing conditions. Wholesalers ship to garden centers around the country, or even in other countries. In general, local garden centers are more likely to buy what they can from local sources, while large chain stores buy mostly from national sources. Ask your garden center's staff for the source of their materials and encourage them to buy locally to reduce greenhouse gas emissions.

Reduce Your Footprint

Locavore may have been the *New Oxford American Dictionary*'s 2007 word of the year, but *carbon neutral* beat it to the punch as the word for 2006. If you can balance the total emissions you are responsible for generating by reducing where possible and offsetting the rest, you will be carbon neutral.

The first step is to estimate your carbon footprint. GreenerChoices.org has links to several online calculators, and you can find more by typing "carbon footprint" into an Internet search engine. Their calculations vary and no one has rated them yet, so try a few. Some ask you to estimate only your home energy consumption, automobile type and use, and air travel. Others are very specific about energy use and include questions about food purchases and lifestyle. My footprint varied widely depending on the calculator, but the more detailed questions revealed an intermediate footprint that is probably closest to the truth. If the site asks you just a few questions and then takes you to a page for a donation, go to another site.

Once I knew the size of my footprint, I began to reduce it by following the many tips in this chapter, buying a more fuel-efficient car, and making other changes in my lifestyle. I also looked into buying carbon offsets, which are generally unregulated contributions to various clean energy or emission abatement projects. These can be purchased from the footprint calculator web pages, which are often sponsored by nonprofit organizations (offset donations to these groups may be tax-deductible in the United States), for-profit businesses, or local utilities. The projects range from local (perhaps in your area) to global and include purchasing and managing forests, investing in wind farms, developing methane abatement programs for livestock farms, carpool matching, optimizing traffic signals, and even creating electrification stations where long-haul truckers can plug in their vehicles during mandated rest stops rather than maintain systems by running the engine. Prices also vary, but figure about $20 per ton of CO_2 emissions, for a total of around $1,000 per year for most households. Carbon offsets should never be used in place of reducing your footprint, but they can help you fund some worthwhile projects while you work on decreasing it.

Final Thoughts

Prior to New Year's Eve 2000, when most people seemed to be concerned about potential computer-created chaos, Brian and I were having an ongoing spirited debate about whether the internal combustion engine or the silicon chip had the bigger effect on the twentieth century. Both have undeniably changed how we live, and there were many arguments for both. Ultimately, we came down on the side of the engine. Factories produced motorized labor-saving devices such as washing machines and tractors. With the increased efficiency in farming, more rural people were able to move to the cities to take factory jobs. This increase in population and the widespread use of cars and trucks changed how we design cities—suddenly urban sprawl was not only happening but desired as post–World War II baby boom families moved to the spacious suburbs. Families became more far-flung as cars and airplanes allowed faster, easier migration to ever-more-distant cities for jobs. The silicon chip, or some other invention not yet dreamed up, may claim the twenty-first century, but in our opinion, the internal combustion engine owns the twentieth.

Unfortunately, all these exciting changes in our standard of living were contingent on the engines' need for fuel—lots of fuel. Now the thing that changed the twentieth century may change the earth forever, if we let it. It is time to stop arguing over whether climate change is happening and take responsibility. It is simply not realistic to think that we can continue to burn fossil fuels at the rate we have over the past several decades. We need to reduce their use and emissions, and we need to reduce the release of other greenhouse gases as well. However, just reducing fuel use will not stop or even reverse climate change. We need to mitigate our actions. If each of us makes a few small changes, together we can make some big changes to the climate.

Guidelines

- Use several online calculators to estimate your family's carbon footprint. Start making changes to reduce it—follow the suggestions for the garden here and consult other sources for information about your home and lifestyle.
- Use your plants to sequester some of the carbon dioxide in the atmosphere. If you have room, plant deciduous or relatively fast-growing trees. If you do not, consider replacing all or part of your lawn with shrubs or other woody plants.
- Soil is even better than plants at holding carbon, so treat it with respect. Do not overtill, which releases carbon dioxide, and limit your use of fertilizers. (See the soil chapter for more information.)
- Use trees to shade the sunny sides of your house and reduce your reliance on air conditioners and fans. Trees can lead to 25 percent less energy use, and deciduous trees will allow sun to come in during the winter, warming the home and reducing heating costs.
- Carefully consider your use of power garden equipment. Electric tools are slightly more fuel-efficient and less polluting than two-stroke engines, but using your own muscles is better for you and the environment. Know how the power in your area is generated and be especially conservative in using electicity generated with fossil fuels.

- Do not use concrete made with portland cement. (You need not replace old portland cement sidewalks or other surfaces.) Use locally available alternatives whose manufacture produces fewer carbon dioxide emissions.
- Become a locavore. Grow as much of your own food as you can, and visit farmers' markets that feature local farmers—it is fun to know the people who grow your food. If there are any community-supported agriculture (CSA) programs in your area, sign up to get fresh local produce delivered to your door.
- Try to purchase landscape plants that are locally grown. At this time the nursery industry does not have a labeling system for locally grown plants, but in the meantime, ask the staff at your garden center where they purchase their plants. Smaller stores are less likely than large chain stores to purchase from large national sources.
- After doing all of these things, recalculate your footprint. Research the options for carbon offsets and consider investing in the ones that make sense to you.

CHAPTER EIGHT

Recycle, Reduce, Reuse, Repurpose

The packaging for a microwavable "microwave" dinner is programmed for a shelf life of maybe six months, a cook time of two minutes and a landfill dead-time of centuries.

—David Wann, *Buzzworm* magazine, November 1990

I inherited some beautiful quilts made by my ancestors in the 1800s. They cut small bits of cloth and sewed them into interesting and intricate patterns. Much of the cloth probably came from old dresses—once they were too worn to wear, they were repurposed into warm quilts for the family or given to peddlers who sold them to paper companies. Each woman probably owned four or five dresses at any time. Ever notice that old houses often have no, or tiny, closets? People had as much as they needed, and what they needed was modest.

Now that we live in a time of prosperity and excess, we tend to acquire what we want rather than what we simply need. Some people have closets the size of my ancestors' bedrooms. We own dozens of outfits, and stores are stuffed with more for us to choose when we are bored with our current selection. What we seem to miss is that our demand for more stuff means more waste—and not just as a manufacturing by-product. According to a report by the European Environment Agency, waste generation in the European Union is linked to economic growth: in times of prosperity, people buy more stuff and toss what they already have. Conversely, municipalities can tell when the economy is bad: garbage collection goes down as we buy less and keep things longer.

Along with all this excess has come a disregard for, and ignorance of, our waste. You know that when you throw something into a waste bin, it will eventually make it out to the street to be picked up by sanitation workers. Stop right now and answer this: where does it go from there? How does your community treat solid waste? Where does that old milk container, empty nail polish bottle, or apple core go? We have developed an out-of-sight-out-of-mind approach to our garbage that is pretty shocking, especially when you consider that, according to the Environmental Protection Agency, the average person in the United States or Canada generates four to five pounds of garbage *per day*. This is the equivalent of eighty-two thousand football fields covered in six feet of compacted garbage every year. It does not even include all the materials that went into making the goods we ultimately throw away—according to *Natural Capitalism* by Paul Hawken, Amory Lovins, and L. Hunter Lovins, for every one hundred pounds of product on store shelves there are thirty-two hundred pounds of waste generated in associated industrial, mining, and energy production.

Globally we are running out of places to put all that garbage: for instance, in November 2007 the BBC reported a U.K. Local Government Association warning that the country would be out of landfill space in less than nine years unless people substantially changed their practices. The United Kingdom has twenty-seven million gardeners, and a reduction in their waste stream could make a big difference to that landfill space. Hawaii is likewise running out of room for its garbage and is investigating hauling it more than two thousand miles away to the U.S. mainland. Experts from larger landmasses say that fewer, but very large, landfills can still handle plenty of waste, but they acknowledge that these large landfills, even carefully managed, can cause environmental problems by producing excess methane through decomposition and attracting pest species. In the past we have done crazy things with our garbage, filling wetlands with it or tossing it down abandoned mine shafts, but landfills today are a technology-based business, with careful geological siting, state-of-the-art liners to limit groundwater pollution, and monitoring for leaks. Some municipalities have proposed burning garbage to generate electricity—at one point there were more than one hundred incineration plants proposed for California alone—but even if this moves us away from using landfills, what will we do with the toxic ash that it produces?

According to the Environmental Literacy Council, in 2006 the average landfill contained 26 percent paper; 18 percent food scraps; 16 percent plastic; 9 percent rubber, leather, and other textiles; 7 percent each yard waste, metal, and wood; and 6 percent glass. Many of these materials are compostable or recyclable (the EPA estimates that about 67 percent of American garbage could be composted), and it is tragic that they have ended up in landfills. Even if we find better ways to deal with garbage than heaping it in landfills, we have to change this lifestyle. We must act in a way that future generations will respect: recycle as much as we can, reduce the amount of garbage produced, reuse some materials, and repurpose those we do not reuse. Gardeners can do our part.

Recycle: New Life for Old Material

The idea of recycling is very attractive: products we no longer need are turned into raw materials from which other things can be made. Gardeners do this in their own yards when they compost, and we are all familiar with industrial-scale recycling. For instance, newsprint, cardboard, and other paper can be repulped and made into different types of paper. Starting with the product made directly from trees, paper can be recycled four to seven times; repeated recycling leads to shorter fibers with different paper properties. Aluminum, steel, plastic, glass, and many other common products are also commonly recycled. The biggest factor in whether a material is recycled is whether there is a market for the new product that makes the effort of collecting and recycling profitable. Consumers can help by purchasing products made with recycled materials, which will prompt manufacturers to increase the supply. You may be able to recycle more than you think, too. Go to www.earth911.com or call 1-800-CLEANUP to find out what can be recycled in your area and where you should take it.

Composting

Home composting is not difficult, and many find it rewarding. Most organic materials can be used, including leaves, grass clippings, seedless weeds, tea and coffee, eggshells, and vegetable trimmings. However, if you use herbicides, such as a "weed and feed," on your lawn, do not compost the clippings from at least the first two mowing cycles. Some fruits are all right, but stronger-smelling ones may attract pests and are best given to commercial composters unless you have a special food composting system. Do not include unshredded woody pieces that will decay slowly, diseased plants or leaves, animal waste, or meat or milk products. Always exclude the roots of rhizomatous weeds, as they may begin to grow in the fertile soil of the compost bin and make it unusable—I found that out after my first attempt at composting included the roots of hedge bindweed *(Calystegia sepium)*. I had to excavate the whole thing and nearly treat it like hazardous waste.

Good compost is about finding the right carbon-to-nitrogen ratio. Dead leaves and straw, sometimes called the brown material, add carbon. Nitrogen is the "green" part—grass clippings, fresh leaves, green weeds. You want about one part green to two parts brown. Too much green, and it will decompose too fast; too much brown, and it will be too slow. Your compost should be about as moist as a wrung-out sponge. Aerate it by turning or fluffing on a regular basis, and you will have the perfect conditions for microoganisms or worms (no need to add them—they will find it) to turn it into rich, crumbly black gold in several months. If you find it is not breaking down fast enough, try adding some soil from your garden to ensure it has the microbes it needs.

You can do this in a simple pile in a corner of the garden or in any of a variety of commercially available composting bins. The bins have the advantage of containing the compost, presenting a more pleasant appearance, and controlling odors that might attract pests such as rodents. Some are tumblers that are easy to turn and aerate, and some are black, to heat the material more quickly.

Many communities promote composting by contracting with companies to collect and process yard and food waste. Some of these businesses offer curbside pickup of waste, while others require that you bring it to a central collection place. Check with your local solid waste management authorities to see if they have set up a program like this.

If you enjoy composting, you might check whether your community has a Master Gardener program that trains Master Composters. In any case, you can spread the composting message to others.

Those Ubiquitous Plastic Pots

One of the commonest recyclable materials in gardens is plastic, most often found in plastic nursery containers. From the small, flimsy six-packs of annuals to the larger, sturdy black containers for shrubs and trees, most nursery plants come in plastic pots. This was not always true. In the 1950s, the first large wholesale company moved from the then-commonplace clay pot to the plastic pot. By the 1970s, the clay pot was rare and plastic was king. Plastic is advantageous for many reasons, including its cost to make and ship (it is lighter than clay), decreased disease spread, and relative indestructibility. Today, every gardener has a collection of black plastic pots in the garage, and botanic gardens despair when they find stacks left anonymously like unwanted pets at their gates overnight.

There are dozens of resins used to make plastic. Six are commonplace in consumer products, but only two of these are widely recycled—polyethylene terephthalate (PET), used in soda bottles and similar containers and noted with a #1, and high-density polyethylene (HDPE), used in milk jugs and the like and noted with a #2. Other plastics are less recyclable, and many curbside recycling programs will not accept more than PETs or HDPEs. You should know what your area's program will take and should not burden it with plastics it cannot handle.

James Garthe, an agricultural engineer and instructor at Pennsylvania State University's Center for Plasticulture (really), estimates that only 10 percent of the plastic used in agriculture and horticulture is recycled, compared to about 25 percent of soda bottles and milk jugs. One problem is the choice of resins: polypropylene (#5), found in plastic mulch mats, coveralls, and outdoor rugs, and polystyrene (#6), found in bedding plant trays, are much used but also hard to recycle. However, even the #2 pots are rarely recycled.

There are regional efforts to combat this trend. In 2007 the Missouri Botanical Garden began a novel project, with the assistance of local Master Gardeners and financial support from Monrovia Nursery, to keep one hundred thousand pounds of plastic nursery pots from the landfill. Several garden centers had trailers into which gardeners sorted their pots by resin type, aided by Master Gardeners when necessary. (The garden centers were happy to host because the people who came often stayed and shopped.) The plastic was then shredded and used to make landscape timbers for such things as raised vegetable beds. The project was supposed to last one year, but it was so successful that it is still ongoing. While programs like this are uncommon in most communities, more may be started as markets develop for polyethylene. Ask your solid waste department, local botanic gardens, and Master Gardeners to explore creating one. Seattle has just started accepting some types of nursery pots in its curbside recycling—other cities may follow suit.

Why don't nurseries just reuse the pots? The pots would need to be cleaned sufficiently to prevent disease spread, but although industrial-scale cleaners and steamers could do this, they are not being used for the task. Some garden centers do provide space for pot exchanges: bring in your clean, unwanted six-inch-wide or larger plastic pots (the smaller, flimsy bedding pots are not recyclable and mostly not reusable), and small local nurseries or gardeners who want to share potted plants with friends can select what they want. Check with local centers, and if none have pot exchange programs, encourage them to start one.

Reduce and Reuse

While recycling definitely helps with the waste problem, it is far better not to have the materials to deal with in the first place. Some things, such as plants, are easily and fully recyclable through composting at a small enough cost to be economically feasible. Paper recycling technology is well established, and there are many recycled-paper products available. But recycling has costs—the materials must be transported, cleaned, and processed—so simply reusing when possible is even better. Rather than buy something, thinking it could be recycled later, consider whether you really need it and whether you already have something that can do the job. Also remember that some materials cannot be recycled, either because of their composition or because there is not enough demand to balance the cost.

It would be preferable if we did not have the dilemma of deciding what can or should be recycled. According to Elizabeth Royte, writing in *Garbage Land,* packaging makes up 35 percent of household garbage weight in the United States. Garden products (though not necessarily plants) are often overpackaged. Let garden center staff know if this is an issue with their products so they can pass the information on to their suppliers. Buy bulk items that are packaged together when you can, and when buying less, avoid unnecessary extra materials: do you really need a cardboard flat to carry a plant or two? Many of us commonly carry reusable bags to the grocery store. Why doesn't some smart entrepreneur make pretty, reusable flats for gardeners?

Polluter Pays?

A company encases a small pair of pruners in a hard shell of unrecyclable plastic that the consumer must pay to include in her trash and society must pay to house for centuries in a landfill. A nursery uses a pot made of cheap but unrecyclable resin that the gardener, with no choice in his community, throws in the trash. Can we make companies accountable for saddling consumers with more garbage?

Pay-as-you-throw systems are increasingly popular as a way to decrease garbage and increase recycling. Garbage collection rates go up steeply as you put more and bigger bins out on the curb, while yard and food waste and recycling are picked up for free or at a cheaper rate. This is an incentive to recycle, but it can also lead people to dump garbage in parks or in store trash receptacles to avoid paying, with the costs passed on to taxpayers and consumers.

Germany, however, has taken a different approach: because companies choose the packaging for their products, the government holds them responsible for recycling it. In response to a stringent law passed in 1991 that requires them to recycle up to 70 percent of their packaging or face huge fines, a number of companies came together to form a consortium, the Duales System Deutschland (DSD). Participants mark their products with a special Green Dot symbol and pay waste companies for curbside pickup. The DSD then recycles or reuses them. While this system suffers from some of the same problems as consumer-based recycling—notably, more plastic than can be recycled with current technology—it is successful in reducing the amount of waste going into landfills, if mostly by forcing companies to reduce their use of nonrecycable materials, and has been replicated in other countries.

The amount and potential recyclability of packaging materials are the responsibility of the manufacturers of consumer goods, but unless we hold them accountable, either directly as in Germany or by not buying their goods, they will not stop their wasteful practices.

Alternatives to Plastic Pots and Labels

One way to eliminate plastic packaging is to grow woody plants in fields rather than pots. The plants are dug out in spring or fall, and because the root ball is wrapped in burlap, they are called balled and burlapped, or B&B. These plants are sold in the same manner as container-grown plants, and they may be less expensive. The practice was formerly more widespread, but B&B plants, as well as bare-root offerings, are still common many places in the spring. One environmental disadvantage to B&B is that considerable organic material is removed from the field in the process, leading to degradation of the soil over time. There are no disadvantages to buying bare-root plants: they are less expensive to transport, do not use plastic pots, and will establish in your garden faster, thanks to their lack of nursery soil. Just be sure to plant promptly after bringing them home so they do not dry out.

For plants that need containers, decomposable products might be the answer. This can also be a good use of waste products, such as rice hulls, coir, and cow manure. So far decomposable pots are not viable on a large commercial scale, but that may change with new technological advances. These containers are especially promising for smaller bedding plants, which are usually put in unrecyclable plastic. Growers have used pots made of heavy paper, and these have great possibilities—they can even hold up against several months of rain. While relatively sturdy, decomposable pots do not last more than a season or two, after which you can compost them. Some municipalities even allow them in curbside yard waste for composting.

Some companies are producing alternatives to plastic labels and tags, using biodegradable materials such as corn and recycled paper. Like the pots, these labels are not commonly used, but with consumer support they may become more common.

Freecycling

Even the Internet can help you avoid the costs of recycling. Craigslist (www.craigslist.org), started in San Francisco in 1995 by Craig Newmark as a sort of classified list for friends, now serves seven hundred cities in seventy countries and has more than forty million users each month. Have some unused mulch taking up space? Get rid of it quickly by listing it in the "free" secion of your city's web page. One gardener changing her driveway to permeable surface saved a huge amount in hauling fees by offering her jack-hammered concrete chunks as freecycle—and they were gone in a day. If your community does not have a Craigslist site, see if any other groups that serve a similar function are active in your area. Find members of Freecycle Network at www.freecycle.org—they are active in many countries, and all you have to do is get a free membership to post about what you need or want

Batteries

Many garden tools use rechargeable batteries, which have a longer life and so stay out of landfills longer and do not need to be replaced as often. Such batteries may contain heavy metals such as cadmium, mercury, lead, and lithium, which can be toxic if improperly disposed of. Local government can tell you where to take both regular and rechargeable batteries when they no longer hold a charge. Some stores will also take them back for recycling.

to get rid of. You can also put up notices on garden center, office, and other bulletin boards or tweet or e-mail your offers to friends. Many charities also accept donations of reusable materials, and some will pick them up.

Repurpose: Get Creative

Here is the fun part of reducing waste—finding new uses for old objects. The opportunities are almost limitless. Perhaps after reading the chapter on water stewardship you were motivated to replace your concrete driveway with permeable paving. Instead of sending the jackhammered chunks to a landfill, use them (instead of flagstone) to create a terrace set in builder's sand or stack them to create a short wall. If you do not want to use them yourself, find someone who does.

Gardeners make planters out of old bed frames, claw-foot bathtubs, toilets, wood chairs, and high-heeled shoes. Look around the house—as long as it can hold soil and has some drainage, it can be a flowerpot. Now head out to the garden center to find some appropriate plants.

Most of us have reused egg cartons to start seeds. Why not use the rinds of hollowed citrus fruits as well? Cut off the top, and then, after eating the fruit, remove all remaining pulp, poke in a drainage hole, add soil and seeds, and plant. What other kitchen items can be reused?

Other objects from the house can migrate outdoors. Most gardeners have some old sheets and blankets that they use to wrap tender plants in the winter. Perhaps they could also serve as water-permeable weed cloths covered by wood mulch. Hang an empty picture frame from branches to frame a view or an especially noteworthy plant from a vantage point—or take several picture frames apart and organize the pieces into an artistic trellis. Broken dishes can make an attractive mosaic embedded in walls or walks.

Old garden equipment can serve new purposes too. A ladder missing a rung might serve as a trellis for vines. Rusty hand tools can be buried handle-side down to create a unique garden border and hose guide.

The imagination can run wild with inventive ideas. The next time you go to put something in the junk pile, stop and allow it a few moments to inspire you. Have fun!

Final Thoughts

In 1997 a southern California sailor on his way home from a sailing race to Hawaii discovered something that had only been predicted before: the Great Pacific Garbage Patch. Directed by a perfect storm of currents and airflow, plastic garbage forms an island that some claim is twice the size of the continental United States, though it is difficult to determine the patch's boundary. Captain Charles Moore must have thought he was dreaming when he saw his boat surrounded by plastic bags, motor oil jugs, tires, toys, and even traffic cones! It took him a week to sail through this horror, for which we must all claim some responsibility.

Humans are not alone in creating garbage. My mountain beavers clean their tunnels by assiduously pushing out the parts of my garden plants they do not consume. But while decomposers in the soil will handle this trash in a matter of months or less, we humans create huge amounts of waste that is not biodegradable. We need to take responsibility for our waste, including that from the garden, and change our priorities: first, buy less, and second, buy recycled materials, in turn recycling them when we are done.

I know where my waste goes. My garbage is picked up by sanitation workers every Wednesday, taken to a transfer station in North Seattle, and packed into a transportation container that is loaded onto a train in a rail yard in South Seattle. From there it heads south to an arid part of northern Oregon, where it enters a landfill. There it will sit, eventually buried, for perhaps hundreds of years. My yard waste and most of my food waste go into a container that is also picked up every Wednesday and sent north to a private company that contracts with the city to quickly compost waste with state-of-the-art technology. Twice a month the recycling is taken from my curbside container and goes to a sorting center south of town, which sells off the various items to processors and manufacturers.

Is knowing where my waste goes the same as taking responsibility for it? Not if I am still careless in what I produce. I may be able to deal with it more responsibly than some because my progressive community gives me many easy opportunities to recycle and compost. Ultimately, though, it is my responsibility to first reduce my waste, to stop creating garbage, then to recycle, reuse, and repurpose what I can.

Guidelines

- Carefully consider what you buy. If it has more packaging than necessary, rethink purchasing it. Whether you buy it or not, let the merchant know you do not like the packaging and consider informing the manufacturer.
- Purchase goods made partially or completely with recycled materials. Many materials are recyclable, but until there is sufficient demand, not many products will be made from them.
- Practice freecycling—if you cannot reuse or repurpose something yourself, give it to charity or offer to let someone pick it up for free.
- Never throw away one-gallon (six-inch-diameter) or larger plastic nursery pots. If they cannot be recycled in your area, find a local garden or community center that hosts or is willing to host a pot exchange, and get garden writers to publicize the event.
- Never burden recyclers with plastics that cannot be recycled.
- Reduce the demand for plastic pots—and maybe save a little money—by purchasing B&B (balled and burlapped) and bare-root plants in the spring.
- Encourage local garden centers to buy bedding plants in decomposable pots if possible. If they already do this, tell them how glad you are.

- Compost yard waste, either through a community program or by yourself. Home composting does not require equipment: leaves, grass clippings, water, and air are the critical ingredients.
- You have heard of locavores (see climate change chapter)—become a loca-reuser. Have fun repurposing things that have outlived their original function. Your imagination is the limit!

EPILOGUE

Toward a Garden Ethic

All ethics so far evolved rest upon a single premise: that the individual is a member of a community of interdependent parts. His instincts prompt him to compete for his place in that community, but his ethics prompt him also to co-operate (perhaps in order that there may be a place to compete for).

—Aldo Leopold, *A Sand County Almanac*

Since Aldo Leopold wrote those words in 1949, his philosophy has been absorbed slowly but increasingly into how we view the protection of natural areas. Building on the work of Leopold, Rachel Carson, and others, we have enacted many local, national, and international laws and treaties that aim to protect the "land"—the natural land and water communities. We still struggle with maintaining it for its intrinsic worth, as opposed to the economic value of the resources and services it provides, but the direction we are taking is positive.

A garden ethic reflects those values—that the gardens we nurture provide food for the soul as well as the table, and that we have a responsibility to manage them in a way that reflects their connections to other gardens and to wildlands. Most gardeners sincerely want to do what is best for the land—we want to support wildlife in our gardens and we want healthy water and soil in our communities—and many common gardening practices are inherently conscientious. You cannot get much more sustainable in a garden than composting green waste or reusing rainwater through barrels or cisterns. You may also already use mulch to conserve water and reduce weeds. Many of my suggestions, such as creating layers of foliage to slow rainfall and prevent stormwater runoff or planting trees near the house to reduce energy use, are simple steps. Others may require more effort. I generally find resistance to eliminating lawns, for example, and many people just cannot bear to reduce or eliminate soil amendments for woody plants. I hope you feel empowered to take the steps that are right for you now and will consider additional conscientious choices when they seem appropriate. Every such choice you make moves us toward a more sustainable future.

As Peter Raven notes in the foreword, the global human population is increasing at a staggering rate. But our burgeoning numbers do not mean that we have to have a negative effect on the earth. E. O. Wilson, the renowned conservation biologist and expert on ants, makes the point that ants have the greatest total biological mass of any animal family on the planet—including humans. They move even more soil than earthworms, so they have an impact on the ecosystems where they are present. Nevertheless, they live in balance with the earth. We can learn to do that too.

Many people making small changes can make a huge collective change. Take Seattle: the city amended its building codes to require water-efficient toilets in new and remodeled properties, and it priced water higher to discourage use. The Seattle Public Utilities office also gave a low-flow showerhead to everyone who returned a mailed coupon. Despite increasing the number of people it services, the utility now delivers twenty million fewer gallons of water per day than it previously did. What if all gardeners stopped or reduced their fertilizer and pesticide use, carefully selected plants and mulched to reduce water use and to prevent plant invasions, and made a home for wildlife in their garden? Margaret Mead, the famous anthropologist,

Conscientious Choices

As you have gone through this book, you have doubtless seen practices you already employ, practices you started to implement, and practices you did not think would be practical for your garden. Since gardening is a process, some of these practices will change over time. The constant for the conscientious gardener is adapting to and learning from the land, which itself constantly adapts to new conditions, and from science, which continuously adds to our knowledge. As you read through the lists below, consider which points are already part of your Conscientious Credo—and which you would like to add.

Soil

- I know what kind of soil I have and add nutrients only as needed.
- I choose plants that will do well in my native soil rather than trying to change my soil with amendments.
- I avoid or limit the use of *Sphagnum* peat moss.
- I use mulch on all my planting beds to feed the soil, suppress weeds, and prevent compaction in high-use areas.
- I use sustainably obtained, preferably locally produced mulch as well as shredded or whole undiseased fallen leaves from my garden.
- I do not dump nonnative worms outside, especially near natural areas.

Water

- I know what watershed I live in.
- I limit access to natural streams or ponds on my property to prevent erosion and to protect animals in the water.
- I never introduce nonnative plants or animals to natural water bodies.
- I do not use more fertilizers or pesticides than absolutely needed, and I never apply them before a storm.
- I do not use water to clean debris from pavement.
- I slow the movement of water so it is absorbed into the soil by using plant layers, bioswales, rain gardens, green roofs, or other methods.
- I use water-pervious materials wherever possible.
- I have a rain barrel or cistern to collect the water from my roof and other impervious surfaces to reuse for irrigation.
- I select plants that use less water and add mulch around them to reduce evaporation. I irrigate only in the absence of adequate precipition and use efficient methods such as drip irrigation.
- I have minimized my lawn or replaced it with ecoturf suited to my climate.

Native Species

- I choose plants that are suited to my climate, soil, and other conditions, using native species when appropriate.
- I have planted native plants that native wildlife can use.
- If I have any doubts, I ask my garden center where its products came from. I look for labels that say the plant was propagated in a nursery or is from cultivated rather than wild stock.
- I have investigated which native species grow wild near my garden in parks, greenbelts, and natural areas. I have taken care not to plant nonnative genotypes (those from plants more than one hundred miles away) near them.
- I do not use rare native species in my garden, especially cultivars, unless I know they are not found elsewhere near my house.
- I have joined a local native plant society.
- When I have time, I volunteer to work with groups doing native plant salvage and restoration projects and native rare plant recovery.

Invasive Species

- I know which garden plant species invade in my region (see the appendix for a list) and check with my local extension agent or state or county weed control board if I am unsure.
- I keep an eye out for and remove wildland invaders from my property.
- I never plant invasives, but if one is appealing, I find a noninvasive alternative with similar traits.
- I do not trade plants with other gardeners if I know the plants are invasive.
- I let garden center employees, catalog suppliers, and Internet sources know I want only noninvasive species.
- I ask garden writers and public gardens to not promote or display species that invade in my area.
- I report species that I think might be a problem to local herbaria or agencies such as a department of agriculture for identification and reporting.

Wildlife

- I have learned about native wildlife that are good garden inhabitants and have made my garden attractive to them.
- I have created structural diversity for shelter and nesting, including shrub and tree layers.
- I have added a water feature, such as a birdbath or small pond.
- I have made my garden safe for pollinating insects by limiting my insecticide use and providing food.
- I tolerate some damage from wildlife, including insects.
- I never relocate animals into my garden but let them find it naturally.
- I never trap and release wildlife from my garden into new areas.
- I have animal-proofed my house to prevent the need to kill or remove wildlife.

Pest Control

- I practice Integrated Pest Management.
- I keep a journal of pest damage to track problems before I begin control measures.
- I practice good sanitation to prevent disease spread.
- I use pesticides only when all other steps have failed and I cannot tolerate the pest. I use them strictly according to the label directions and store and dispose of them safely.
- I prevent seed production by weeds, removing either the plants or the seed heads before seeds spread.
- I dispose of weeds appropriately, according to the applicable rules of my local government.
- I have eliminated plants that are persistent pest attracters and replaced them with pest-resistant species.

Climate

- I have used online calculators to estimate my family's carbon footprint. I have made changes to reduce my footprint and invested in carbon offset projects to balance the rest.
- I have maximized the amount of carbon I can store in my garden by planting some deciduous trees and shrubs and not tilling the soil more than necessary.
- I have planted trees to shade the sunny sides of my house to reduce energy use.
- I use hand tools when possible and electric when needed, and I minimize the use of two-stroke engines.
- I grow as much of my own food as I can and visit farmers' markets that feature local farmers to reduce the greenhouse gas emissions associated with shipping.
- I purchase locally grown landscape plants whenever possible.
- I have let my garden center know I prefer to buy locally produced plants.

Waste Disposal and Recycling

- I do not buy products that are overpackaged.
- I have told merchants and manufacturers that I will not buy their overpackaged products.
- I purchase goods made partially or completely with recycled materials.
- I am a loca-reuser: I enjoy finding new purposes for old objects.
- I never throw away one-gallon (six-inch-diameter) or larger plastic nursery pots—I reuse them or give them to a nursery or business that will reuse or recycle them.
- I minimize my need for plastic pots by purchasing bare-root and balled-and-burlapped (B&B) plants in the spring or growing them myself from seeds or cuttings.
- I encourage garden centers to purchase plants grown in decomposable pots.
- I practice yard waste composting through a community program or by myself.
- When appropriate I practice freecycling.

once said, "Never doubt that a small group of thoughtful, committed citizens can change the world. Indeed, it's the only thing that ever has." It is the power of one . . . plus one . . . plus one . . .

It has really hit home with me in the course of researching and writing this book that our mobile society has become disconnected from its surroundings. Do you know what native plant species are found near your house? Do you know what garden species invade wildlands in your region? Do you know your watershed? Raise your hand if you really know what type of soil you have. Your ancestors—perhaps many years back, perhaps more recently—lived in one place for generations and knew the land. They understood the interconnectedness of soil, plants, and animals. As the great conservationist John Muir said, "When one tugs at a single thing in nature, he finds it attached to the rest of the world." The conscientious gardener, like his or her conscientious forebears, understands, protects, and promotes those connections.

This work is not possible without some effort, especially as more and more companies engage in greenwashing, making claims of environmental sustainability that are nothing more than claims. When confronted with such assertions, use your common sense, and if something doesn't seem right, ask questions. Ask questions anyway—that is how we learn!

Wouldn't it be nice if there were some neutral third party that could make these checks for us, certifying which nurseries or landscape designers make decisions with Earth, and even fair treatment of their workers, in mind? Then you could use services and products you know to be conservation-friendly without having to investigate them yourself. Half the work toward following your garden ethic would be done. That would be wonderful! Therefore, it is with great sadness that I must say that such a thing does not exist—yet. While there is interest in developing third-party certification programs for nurseries, we are far from such a system. There are some efforts under way, however, that attempt to address environmental concerns.

The VeriFlora program (www.veriflora.org) was started by people in the floral industry who were concerned that cut flowers were being produced in unsustainable ways and that employees, often in developing countries, were not being treated fairly. Working with Scientific Certification Systems (www.scscertified.com), it has developed standards that ensure that many, though not all, of the concerns (such as reducing invasive plant introductions) I have raised in this book are addressed. The standards are still being confirmed, but VeriFlora has started registering some companies that have been approved. Most are in the floral business, but VeriFlora looks forward to expanding to more nurseries in the coming years.

The Sustainable Sites Initiative (www.sustainablesites.org), a program of several botanic gardens and the American Society of Landscape Architects, aims to give architects, designers, and landscapers good information about developing sustainable gardens. These might be parks, urban open spaces, or landscapes around buildings. By spring 2011 it plans to have a ranking system that will objectively rate landscapes according to strict conservation criteria.

In Australia a group of garden centers has come together to start a non-profit organization called Sustainable Gardening Australia (www.sgaonline.org.au). They "envisage a world where people cultivate their gardens in sympathy with the environment by following the seasonal rhythms of the land, its ecology and climate." While not a third-party certifier, the group requires nurseries to put their staff through a rigorous training program and submit to independent audits to maintain membership. The website has helpful fact sheets and a list of member public gardens, garden centers, and landscapers.

Aldo Leopold noted that he "presented the land ethic as a product of social evolution": as new discoveries about the benefits or harms of land or wildlife management practices were realized, acceptance of those practices was modified. I also believe that the garden ethic will evolve as we understand more about our world. I have provided some ideas toward achieving it—giving context and guidelines to help you understand the consequences of your actions—but each individual must choose a personal path of right and wrong. The next steps are up to you. I look forward to the future that conscientious gardeners will create.

I began the book with a quote from Leopold, and I will close with perhaps the most iconic one from the same essay, "The Land Ethic," from *A Sand County Almanac*. It says all we need to know about ethics and how to choose right from wrong. "A thing is right when it tends to preserve the integrity, stability, and beauty of the biotic community. It is wrong when it tends otherwise."

ACKNOWLEDGMENTS

A small group of thoughtful and committed citizens inspired me to write this book. The faculty, staff, and students at the University of Washington Botanic Gardens have provided intellectual challenge and stimulating discussions for many years. Graduate students Wendy DesCamp, Jeff Richardson, Doug Schmitt, Sierra Smith, Patrick Swartzkopf, and Julia Tracy helped me frame the issues, and their comments stayed with me as I wrote. Dave Hays stepped in to modify Figure 3. Linda Chalker-Scott, with her intolerance for poor horticultural science, provided stimulating conversations; Kern Ewing, his humorous, insightful, and sometimes frustratingly correct viewpoint. Jenn Leach generously allowed me to use her photographs on the cover. I benefited from the many years of Integrated Pest Management experience of Charlie Shull from Sky Nursery in Seattle. Wendy DesCamp, Bob Edmonds, Soo-Hyung Kim, and Karen Preuss read portions of the book and provided valuable feedback. Jim Affolter, Sierra Smith, Astrid Volder, and an anonymous reviewer gave me invaluable comments on the whole manuscript. My editors at University of California Press were amazingly patient and thorough. My mother, Ruth Phillips, gave me a love of botanical science, and my father, Bill Phillips, his love of gardening—I only wish they were still with me. I appreciate that my feline companions, Zoë, Zeke, and Zach, welcomed me into their pride during the sabbatical year I spent at home researching and writing the book. Finally, I thank my amazingly supportive husband, Brian, who has abetted and encouraged my obsession with gardening and my intellectual development as a scientist. I am truly lucky to have him in my life.

APPENDIX

Global List of Invasive Garden Plants

Species	Common Name(s)	Native To
TREES AND SHRUBS		
Acacia auriculiformis	Earleaf acacia	New Guinea, parts of Australia
Acacia baileyana	Cootamundra wattle	Southeast Australia
Acacia confusa	Formosan koa	Asia
Acacia decurrens	Green wattle	Australia
Acacia farnesiana	Mimosa bush, sweet wattle	Neotropics
Acacia melanoxylon	Blackwood acacia	Australia
Acer ginnala	Amur maple	Northern Asia
Acer negundo	Box elder	North America
Acer platanoides	Norway maple	Northern Europe

Impacts	Cautions for Use
Alters soil chemistry by fixing nitrogen, which can put native plants at a disadvantage	Avoid in tropical areas, especially islands; now invading in Florida, the South Pacific
Alters soil chemistry by fixing nitrogen; hybridizes with other acacia	Now invading in Hawaii, South Africa, parts of Australia where it is not native and other native acacia exist
Alters soil chemistry by fixing nitrogen; may be allelopathic	Avoid in tropical areas, especially islands; now invading in Hawaii, Taiwan
Alters soil chemistry by fixing nitrogen; may be allelopathic; research shows it uses more water than natives in South Africa	Avoid in Mediterranean climates; now invading in California, South Africa, parts of Australia where it is not native and other acacia exist
Alters soil chemistry by fixing nitrogen; one of the most invasive acacias	Avoid in dry, warm climates; now invading in South Africa, Australia, Madagascar
Alters soil chemistry by fixing nitrogen; can form root suckers that increase its spread	Avoid in Mediterranean climates; now invading in California, South Africa, and New Zealand
Shades out native species in grasslands and open forests, where native species need light; prolific seeder	Now invading in the East and upper Midwest of the United States
Traps sediment along rivers and streams, causing them to change their course	Now in parts of Australia, central Europe; may invade parts of the United States where it has been introduced, especially along rivers
Shades out understory species in forests	Avoid in temperate areas with summer rainfall; now invading in the United States' Northeast and Mid-Atlantic

Species	Common Name(s)	Native To
TREES AND SHRUBS *(continued)*		
Acer pseudoplatanus	Sycamore maple	Northern Europe
Ailanthus altissima	Tree-of-heaven	Asia
Albizia julibrissin	Silk tree	Southern and eastern Asia
Albizia lebbeck	Woman's tongue	Tropical Asia
Ardisia crenata	Coral berry	Japan to India
Ardisia elliptica	Shoebutton ardisia	India, Malaysia
Bauhinia variegata	Orchid tree	South Asia
Berberis darwinii	Dawin's barberry	Chile, Argentina
Berberis thunbergii	Japanese barberry	Japan
Bischofia javanica	Bishopwood	Tropical Asia
Bocconia frutescens	Bocconia, plume poppy	Neotropics
Broussenetia papyrifera	Paper mulberry	South Pacific
Buddleja asiatica	Dog tail	Asia
Buddleja davidii	Butterfly bush	China

Impacts	Cautions for Use
Invades open wood and competes with native species; can be especially invasive in wetlands	Now invading in northeastern United States, temperate parts of South America, New Zealand, the British Isles
Forms dense stands that exclude native species; allelopathic	Do not use anywhere
Found in disturbed areas and around streams, where it may create more shade	Avoid in temperate climates with high summer rainfall; now invading in southeastern United States
Invading throughout the tropics; rapidly expanding in some places	Avoid in wet tropics; now invading in Florida, the Caribbean, the South Pacific, South Africa
Shade tolerant; replaces native vegetation in forests; can reach densities of up to one hundred plants per square meter	Avoid in wet tropical and subtropical areas; now invading in Florida, Hawaii, Mauritius
Alters soil chemistry by fixing nitrogen; dominates forest understories	Avoid in warm, wet areas; now invading in Hawaii, Florida, parts of the Caribbean, Indonesia
Displaces native vegetation in at least two different plant community types in Florida	Now invading in Florida, parts of the Caribbean, some parts of the South Pacific, South Africa
Invades several plant community types without the aid of prior disturbance; forms dense colonies	Avoid in mild temperate climates; now invading in New Zealand, Australia
Can form dense stands that exclude native species	Avoid in temperate areas with cold winters; now invading in the northeastern United States
Invades several different native plant communities in Florida	Avoid in wet tropics; now invading in Florida, the South Pacific
Forms dense stands in Hawaii, excluding native species; prolific seeder; difficult to control	Avoid in dry tropics; use cautiously in the wet tropics; now invading in moderately moist forests in Hawaii
Forms dense thickets; decreases wildlife habitats	Avoid in mild areas with high summer rainfall; now invading in the southeastern United States
Invades disturbed areas, altering forest succession	Avoid in warm climates; now invading in the South Pacific
Changes sediment trapping in streams and rivers	Use with care everywhere; now invading in parts of the United States, Great Britain, France, New Zealand

Species	Common Name(s)	Native To
TREES AND SHRUBS *(continued)*		
Buddleja madagascariensis	Smoke bush	Madagascar
Caesalpinia decapetala	Mysore thorn, wait-a-bit	Tropical India
Calophyllum antillanum	Alexandrian laurel, Santa Maria	Caribbean islands
Calotropis procera	Calotrope, rubber bush	India to northern Africa
Casuarina equisetifolia	Ironwood, Australian pine	Australia, southern Asia, South Pacific islands
Casuarina glauca	Suckering Australian pine	Australia
Celtis australis	Nettle tree, Mediterranean hackberry	Mediterranean region
Celtis sinensis	Chinese elm, hackberry	China
Cestrum diurnum	Day jessamine	Tropical America
Cestrum nocturnum	Night jessamine	Mexico, West Indies
Cestrum parqui	Green cestrum	Chile
Chrysanthemoides monilifera ssp. monilifera	Boneseed, bitou bush	South Africa
Cinnamomum camphora	Camphor tree	Asia
Citharexylum spinosum	Fiddlewood	West Indies

Impacts	Cautions for Use
Forms dense thickets; excludes native plants	Avoid in moist, wet areas of warm climates; now invading in Hawaii, Australia, South Africa
Dense thorns impact wildlife and livestock and make plant difficult to control	Now invading in Hawaii, other tropical areas
Invades mangrove and other coastal forests, where it can outcompete native species	Avoid in warm coastal areas, especially near mangrove forests
Very competitive with native vegetation	Now invading in Hawaii, Australia, other islands
Forms dense stands producing deep litter that smothers and poisons nearby plants; alters soil chemistry by fixing nitrogen	Do not use anywhere outside of native range
Affects plant communities as above for *C. equisetifolia;* also known to replace nesting sites for American crocodiles and turtles	Do not use anywhere outside of native range
Invasive roots that may allow it to outcompete native species and lift pavement	Now invading in parts of Australia, especially around Canberra; use with caution in other areas
Outcompetes native vegetation	Now invading in South East Queensland in Australia; use with caution in other areas
Outcompetes native vegetation; fruit poisonous to mammals	Avoid in warm, wet areas; now invading in parts of the southern United States, Hawaii, Puerto Rico, Guam, American Samoa
Forms dense, impenetrable thickets in wet forests in tropical climates	Avoid in warm, wet areas; now invading in southern United States, Hawaii, many other islands in the South Pacific
Invades along rivers, where it is very competitive with native plants	Avoid in warm, wet areas; now invading in Hawaii, other South Pacific islands, Australia, New Zealand
Decreases native species abundance	Avoid in Mediterranean climates; now invading in Australia, New Zealand
Forms dense thickets; outcompetes native vegetation	Avoid in warm areas with abundant rainfall; now invading in southern United States, Hawaii, parts of Australia
Grows fast; produces seeds at an early age; quickly dominates	Now invading in parts of the southern United States, Hawaii

Species	Common Name(s)	Native To
TREES AND SHRUBS *(continued)*		
Clerodendrum philippinum	Pikake, glory bower	South China
Clidemia hirta	Koster's curse	Neotropics
Coprosma repens	Mirror bush	New Zealand
Cotoneaster franchetti	Grey cotoneaster	Tibet, China
Cotoneaster glaucophyllus	Largeleaf cotoneaster	China
Cotoneaster pannosus	Silverleaf cotoneaster	China
Crataegus monogyna	English hawthorne	Europe, northwestern Africa, western Asia
Cupaniopsis anacardioides	Carrotwood	Australia
Cynanchum louiseae*, syn. *Vincetoxicum nigrum	Black swallow wort	Southern Europe
Cytisus scoparius	Scotch broom, Scot's broom	England, continental Europe
Cytisus striatus	Portuguese broom	Southern Europe
Dalbergia sissoo	Indian rosewood	India
Daphne laureola	Spurge laurel	Southern Europe

Impacts	Cautions for Use
Can be very aggressive along roadsides; spreads into native vegetation; aggressively sprouts roots	Now invading in tropical South America, Africa; do not use in tropical areas unless roots are restricted
Forms dense, impenetrable thickets of shrubs in sunny, open areas	Do not use anywhere outside of native range
Competes with native plants	Now invading in New Zealand and parts of Australia, such as New South Wales and Tasmania; beginning to spread in California
Grows fast; outcompetes some native vegetation	Now invading in northwestern United States, Tasmania; should be monitored in other Mediterranean climates
Invades along rivers, where it grows rapidly; may also invade in dry upland areas; outcompetes native vegetation	Now invading in New South Wales in Australia, New Zealand
Colonizes wooded areas; can invade dry areas	Now invading in Mediterranean climates such as California, parts of Australia (Tasmania, Victoria, New South Wales)
Fruits favored by birds, decreasing dispersal of native species; may hybridize with native relatives	Avoid in temperate areas with cold winters where native *Crataegus* species are found
Invades many native community types, including coastal forests; outcompetes other aggressive invasive species	Now invading in coastal southern Florida; should be monitored wherever it is grown
Very aggressive shrub/vine; covers and chokes native plant species; replaces a plant species in northeastern United States required for monarch butterfly reproduction	Avoid in temperate areas; now invading in northeastern United States into Missouri—do not use in that region
Alters soil chemistry by fixing nitrogen	Avoid in areas with wet, mild winters, dry summers
Alters soil chemistry by fixing nitrogen	Avoid in areas with wet, mild winters, dry summers
Alters soil chemistry by fixing nitrogen	Avoid in warm, wet climates; now invading in southern Florida, Hawaii, Taiwan
Extensive root system can outcompete native vegetation in forests	Avoid in temperate areas with winter rainfall; now invading in northwestern United States

Species	Common Name(s)	Native To
TREES AND SHRUBS *(continued)*		
Daphne mezereum	February daphne	Europe, western Asia
Elaeagnus angustifolia	Russian-olive	Western and central Asia
Elaeagnus umbellata	Autumn-olive	Eastern Asia
Erica lusitanica	Portuguese heath, Spanish heath	Southern Europe
Eucalyptus globulus	Blue gum	Australia
Euonymus alata	Burning bush, winged spindle bush	Asia
Euonymus fortunei	Wintercreeper euonymus	China
Ficus altissima	False banyan	Tropical Asia
Ficus microcarpa	Chinese banyan	China
Fuchsia magellanica	Hardy fuchsia	Chile
Genista monspessulana	French broom	Europe
Grevillea banksii	Kahili flower	Australia
Grevillea robusta	Silk oak	Australia
Hakea sericea	Needlewood	Australia

Impacts	Cautions for Use
Scattered in moist forest, may not have serious impact	Avoid in cold, moist areas; now invading in Canada, parts of the northeastern United States
Alters soil chemistry by fixing nitrogen; traps sediment along rivers	Now invading in western and northeastern United States; should be monitored near rivers in drier climates
Alters soil chemistry by fixing nitrogen; forms dense, thorny thickets	Now invading in northeastern and midwestern United States; monitor everywhere
Crowds out native plants; alters soil chemistry; produces large amounts of seeds	Now invading in northern coastal California, Australia (Victoria, New South Wales, Tasmania), New Zealand
Copious oils in leaves increase fire danger	Avoid in Mediterranean climates
Invades many community types; forms dense thickets	Avoid in temperate climates with summer rainfall; spreading fast in northeastern and midwestern United States
Fast-growing vine; competes with native species	Avoid in areas with summer rainfall
Spreads in native forests, but impacts are unknown	Now invading in southern Florida, where specialist pollinators have been introduced
Invades native forests; can strangle trees	Now invading where specialist pollinators have been introduced in Hawaii, southern Florida, the Bahamas, parts of Mexico
Forms dense thickets in open woods and forest edges	Avoid in cold, moist areas; now invading in high-elevation Hawaii, Ireland, Réunion, Tasmania
Shades saplings, preventing growth; alters soil chemistry by fixing nitrogen; highly flammable and may increase fire frequency	Avoid in Mediterranean climates; now invading in California, many parts of Australia (including Tasmania)
Allelopathic	Now invading in many parts of Hawaii and KwaZulu-Natal, South Africa
Allelopathic	Now invading in Hawaii and the Caribbean, especially Jamaica
Highly competitive with native plants	Avoid in Mediterranean climates; now invading in South Africa

Species	Common Name(s)	Native To
TREES AND SHRUBS *(continued)*		
Hiptage benghalensis	Hiptage	India, southeastern Asia
Hypericum canariense	Canary Island St. John's wort	Canary Islands
Ilex aquifolium	English holly	England, northern Europe
Jasminum sambac	Arabian jasmine	Southern Asia
Jatropha gossypifolia	Bellyache bush	Caribbean
Koelreuteria elegans	Golden rain tree	Taiwan, Fiji
Lantana camara	Lantana	Central America
Leptospermum laevigatum	Tea tree	Australia
Leycesteria formosa	Himalayan honeysuckle	Himalayas
Ligustrum japonicum	Japanese privet	Japan
Ligustrum lucidum	Broad-leaf privet	Asia
Ligustrum obtusifolium	Border privet	Japan
Ligustrum sinense	Chinese privet	China
Lonicera × bella	Bell's honeysuckle	Hybrid, garden origin

Impacts	Cautions for Use
Smothers native vegetation	Avoid on warm-climate islands; now invading in Hawaii, Florida, Mauritius, Réunion
Grows rapidly; outcompetes native vegetation	Now invading in Calfornia, Hawaii, southwestern Australia
Invades forests	Avoid in cool, moist forests; now invading in northwestern United States
Outcompetes native plants	Avoid in Florida, Hawaii, the Caribbean
Smothers native vegetation; raw seeds poisonous to humans and other animals; sap is an irritant	Avoid in wet tropical areas; now invading in Hawaii, northern and western Australia
Outcompetes native plants	Avoid in wet tropical climates; now invading in Florida, Hawaii
Forms dense, impenetrable thickets with sharp thorns	Avoid in tropical areas
Forms dense thickets that outcompete native plants	Avoid in Mediterranean climates; now invading in Hawaii, South Africa
Forms dense thickets that outcompete native plants	Avoid in cool temperate climates; now invading in New Zealand, Australia (New South Wales)
Forms dense thickets in forests; outcompetes native plants	Avoid in areas with warm summers with regular rainfall, mild winters; now invading in southeastern United States, New Zealand
Invades forests; excludes native vegetation	Avoid in temperate areas with summer rainfall; now invading in eastern United States, Australia (Queensland)
Invades forest edges; grows in dense thickets	Avoid in temperate areas with summer rainfall; now invading in northeastern and midwestern United States
Forms dense thickets in forests; outcompetes native plants	Now invading in eastern United States, Australia (New South Wales); should be monitored in temperate climates
Mainly spreads vegetatively from yard waste disposal; can form large patches that exclude native vegetation	Now invading in eastern and midwestern United States; never dump waste material near wildlands

Species	Common Name(s)	Native To
TREES AND SHRUBS *(continued)*		
Lonicera japonica	Japanese honeysuckle	Japan
Lonicera maackii	Amur honeysuckle	Korea, Manchuria
Lonicera morrowii	Morrow's honeysuckle	Japan
Lonicera tatarica	Tatarian honeysuckle	Central Asia
Lupinus arboreus	Tree lupine	Northern California
Macaranga mappa	Bingabing	Australia, New Zealand, Melanesia
Melaleuca quinquenervia	Paperbark tree, cajeput	Australia
Melia azedarach	China berry	India, China
Miconia calvescens	Miconia	South America
Mimosa pigra	Cat-claw mimosa	Tropical America
Morella faya, **syn. *Myrica faya***	Firetree	Southern Europe
Moringa oleifera	Horseradish tree	India
Nandina domestica	Heavenly bamboo	China, Japan
Nicotiana glauca	Tree tobacco	South America

Impacts	Cautions for Use
Quickly covers shrubs; outcompetes native vegetation	Avoid in areas with summer rainfall; now invading in eastern United States, many South Pacific islands, the Canary Islands, parts of New Zealand, Australia
Forms dense thickets in forests	Avoid in cool, temperate climates; now invading in eastern United States
Invades wet and dry areas in open forests	Avoid in cool, temperate climates; now invading in eastern United States
Allelopathic	Avoid in colder temperate areas; now invading in eastern United States
Alters soil chemistry by fixing nitrogen	Avoid in temperate areas with sandy soil; now invading in northwestern United States, Great Britain
Forms dense thickets that exclude native vegetation	Avoid in wet tropics; now invading in wet areas of Hawaii
Forms dense stands with little room for animal habitat; allelopathic	Avoid near tropical wetlands; now invading in Florida, some islands in the South Pacific
Outcompetes native plants	Use with care everywhere; now invading in southeastern United States, Neotropics, South Africa, South Pacific
Aggressively replaces forests	Do not use anywhere; now invading in Hawaii, Tahiti, South Pacific islands
Obstructs water flow in rivers and streams; decreases wildlife habitat	Do not use anywhere; now invading in Florida, northern Australia, Vietnam
Alters soil chemistry by fixing nitrogen	Now invading in Hawaii; monitor everywhere
Forms dense thickets	Now invading in South Pacific, Caribbean islands
Forms dense thickets; new cultivars may be safer	Now invading in Hawaii, eastern United States; use with caution everywhere
Forms dense stands; leaves are poisonous	Avoid in Mediterranean climates; now invading in California, Australia, Mexico, South Africa, Israel, India

Species	Common Name(s)	Native To
TREES AND SHRUBS *(continued)*		
Ochna serrulata	Ochna, Mickey Mouse plant	South Africa
Paraserianthes lophantha* ssp. *lophantha	Cape wattle	Australia
Paulownia tomentosa	Princess tree, empress tree	Eastern Asia
Pittosporum undulatum	Mock orange, sweet pittosporum	Australia
Polygala myrtifolia	Myrtle leaf milkwort, sweet pea	Africa
Prunus laurocerasus	Cherry laurel	Europe
Prunus lusitanica	Portuguese laurel	Southern Europe
Psidium cattleianum	Strawberry guava	Brazil
Psoralea pinnata	Blue psoralea, blue butterfly bush	South Africa
Rhamnus cathartica	Common buckthorn	Eurasia
Rhapiolepis indica	Indian hawthorn	India
Rhododendron ponticum	Rhododendron	Southeastern Europe, western Asia
Rhodomyrtus tomentosus	Downy rose myrtle	Southeastern Asia
Ricinis communis	Castor bean	Africa

Impacts	Cautions for Use
Invades forests and wildlands	Avoid in warm, wet climates; now invading in several South Pacific islands, Australia (Queensland)
Invades coastal forests	Now invading in Hawaii, parts of Australia where it is not native, South Africa, New Zealand
Outcompetes native vegetation; aggressively spreads by root suckers; each tree can produce twenty million seeds per year	Now invading in eastern United States
Creates a dense forest understory, where it replaces native vegetation	Avoid in warm places where it is cultivated; now invading in California, Jamaica, Hawaii, other Pacific islands, South Africa, parts of Australia where it is not native
Invades forests along rivers	Avoid in temperate areas; now invading in Australia, New Zealand
Can form dense stands in urban forests	Avoid in areas with mild, wet winters; now invading in northwestern United States, Jamaica (Blue Mountains), Australia (New South Wales)
Can invade urban forests	Avoid in areas with mild, wet winters; now invading in northwestern United States, Jamaica (Blue Mountains)
Forms very dense stands; may be allelopathic	Avoid in warm, wet climates; now invading in Florida, many South Pacific islands
Forms dense stands; difficult to control	Now invading in Australia (Tasmania, Victoria, Western Australia), New Zealand; should be monitored in mild temperate climates
Invades grasslands and wet and dry forests	Avoid in temperate areas; now invading in northern United States, Canada
A minor invader; impact is unknown	Now invading in Australia (Queensland)
Forms dense thickets; allelopathic	Now invading in Great Britain, New Zealand
Invades open forests; may increase fire frequency	Now invading in warm parts of the United States (such as Florida, Hawaii), Thailand
Forms dense thickets	Now invading in the tropics; probably safe in temperate areas when used as a containerized annual

Species	Common Name(s)	Native To
TREES AND SHRUBS *(continued)*		
Robinia pseudoacacia	Black locust	Eastern United States
Rosa canina	Dog rose	Europe
Rosa eglanteria*, syn. *Rosa rubiginosa	Sweetbriar	Europe, western Asia
Rosa multiflora	Multiflora rose	Eastern Asia
Rosa rugosa	Rugosa or saltspray rose	China
Sapium sebiferum	Chinese tallow tree	China
Schefflera actinophylla*, syn. *Brassia actinophylla	Octopus tree	Australia, New Guinea
Schinus terebinthifolius	Brazilian pepper tree, Christmas berry	Brazil
Senna alata	Candletree	Neotropics
Senna pendula	Climbing cassia	Tropical South America
Sorbus aucuparia	European mountain ash	Europe
Spartium junceum	Spanish broom	Southern Europe
Spathodea campanulata	African tulip tree	Tropical Africa
Spiraea japonica	Japanese spiraea	Japan
Tamarix aphylla	Athel	Northern Africa

Impacts	Cautions for Use
Alters soil chemistry by fixing nitrogen	Avoid in dry areas along rivers; now invading in western United States, northern Europe
Forms dense thickets	Now invading in much of the United States, southern Australia
Forms dense thickets	Now invading in parts of the United States; should be monitored in temperate climates
Forms dense thickets in grasslands	Now invading in midwestern United States; should be monitored in temperate climates
Forms dense, thorny thickets	Now invading in coastal areas of northeastern and western United States, temperate Europe, southern Asia
Aggressively competes with natives in coastal prairies and wetlands	Avoid in warm temperate areas; now invading in southern and western United States, Australia, South Africa
Invades intact forests; alters understory	Avoid on tropical or subtropical islands; now invading in Florida, South Pacific, and Indian Ocean islands
Forms dense stands; allelopathic	Do not use anywhere; now invading in Florida, Caribbean, South Pacific
Forms dense thickets, especially near water	Avoid in warm areas of United States; now invading in South Pacific, Australia, Africa, Asia
Climbs over plants in forests; displaces native vegetation	Now invading in Hawaii, Australia (Queensland), Bahamas, southwestern Europe; use with caution everywhere
Common in urban forests, but impacts are unknown	Avoid in forests in temperate climates
Alters soil chemistry by fixing nitrogen	Avoid in Mediterranean climates; now invading in western United States, South Africa, Australia
Can form dense thickets in wet tropical forests	Avoid in tropics; now invading in the South Pacific
Displaces native species	Now invading in open forests of United States, central Europe, Australia
Less invasive than *T. ramosissima*, but impacts are unknown	Now invading in southwestern United States, central Australia

Species	Common Name(s)	Native To
TREES AND SHRUBS *(continued)*		
Tamarix ramosissima	Tamarisk	Central Asia, northern Africa
Terminalia catappa	Indian almond	Indonesia
Tournefortia argentea	Tree heliotrope	Tropical Asia, Australia
Ulmus pumila	Siberian elm	Siberia
HERBACEOUS PLANTS		
Aegopodium podagraria	Goutweed	Europe
Ageratina adenophora	Eupatory, crofton weed	Mexico
Arctotheca calendula	Capeweed, South African capeweed	South Africa
Asparagus asparagoides	Bridal veil, bridal creeper, smilax	South Africa
Asparagus densiflorus*, syn. *Asparagus sprengeri	Asparagus fern	South Africa
Asparagus scandens	Climbing asparagus	South Africa
Begonia cucullata	Wax begonia	South America
Centaurea cyanus	Bachelor's button, corn flower	Europe, Near East
Centaurea macrocephala	Bighead knapweed	Caucasus Mountains
Centranthus ruber	Red valerian	Mediterranean
Colocasia esculentum	Elephant ears, wild taro	South Pacific, India

Impacts	Cautions for Use
Has a deep taproot and high water use, reducing the water available to more shallowly rooted native species	Avoid in warm, arid areas; now invading in southwestern United States, South Africa, central Australia
Invades coastal strands	Now invading in Florida, South Pacific, Caribbean
Invades coastal strands	Avoid in tropics; now invading in South Pacific
Forms dense thickets; may hybridize with native elms	Now invading in United States, especially New Mexico and upper Midwest; monitor everywhere
Aggressive rhizomes outcompete native species; hard to control	Can be managed if roots are contained in pots or with root barriers
Poisonous to livestock	Now invading in almost all tropical places where introduced
Scrambles over and competes with plants along rivers and in grasslands and coastal dunes	Avoid in Mediterranean climates; now invading in California, Australia
Invades along rivers and in open forests; smothers and suppresses native vegetation	Now invading in southern California, southern and western Australia, New Zealand
Grows spiny tangles of foliage, which cover plants in dry and moist forests	Now invading in Florida, Hawaii, Caribbean islands, eastern Australia, South Pacific islands
Vine scrambles over surrounding vegetation	Now invading in New Zealand, Australia
Found in moist forests; can form floating mats of debris in streams	Avoid in tropical and subtropical areas; now invading in southeastern United States
Invades grasslands; found in wildflower seed mixes	Avoid near grasslands in North America, Australia
Invades grasslands	Avoid near grasslands; now invading in northwestern and midwestern United States
Mostly invades in urban areas, other disturbed places	Now invading in western United States, high-altitude Hawaii, Australia; monitor near all natural areas
Invades some wetlands	Avoid near wild aquatic systems

Species	Common Name(s)	Native To
HERBACEOUS PLANTS *(continued)*		
Conium maculatum	Hemlock	Europe
Digitalis purpurea	Foxglove	Europe
Dipsacus laciniatus	Teasel	Europe
Echium plantagineum	Viper's bugloss, Paterson's curse	Western and southern Europe
Eschscholzia californica	California poppy	California
Fallopia japonica*, syn. *Polygonum cuspidatum*, *Reynoutria japonica*, also relative *F. sachalinesis* and their hybrid *F. bohemica	Japanese knotweed relatives, giant knotweed, bohemian knotweed	Asia
Foeniculum vulgare	Fennel	Mediterranean
Gomphocarpus fruticosus	Swan plant	South Africa
Gypsophila paniculata	Baby's breath	Mediterranean region into Asia
Hedychium coronarium	White ginger	Himalayas
Hedychium gardnerianum	Kahili ginger	Himalayas
Heracleum mantegazzianum	Giant hogweed	Middle East
Hesperis matronalis	Dame's rocket	Eurasia
Impatiens glandulifera	Policeman's helmet	Himalayas

Impacts	Cautions for Use
Forms dense stands along streams; extremely poisonous	Do not use anywhere; now invading in much of North America, South America, Australia, New Zealand
Poisonous; grows along streams, disturbed areas	Avoid near wildlands
Invades wet areas; has a deep taproot	Avoid near wildlands
Grows densely in open areas; poisons livestock	Now invading in western United States, Australia, Canada
Can be competitive in grasslands	Mediterranean climates; now invading in northwestern United States, Chile
Forms dense stands where nothing else can grow; can affect food webs along rivers; hard to kill	Do not use anywhere outside of native range
Invades grasslands	Form sold for leaves and seeds should not be used; edible form with a bulbous base is not invasive
Forms dense thickets in wet areas	Avoid in Mediterranean climates; now invading in western Australia
Invades along shores of lakes and rivers; competes with some rare plants	Now invading in western United States; monitor everywhere
Invades wet forests	Now invading in South Pacific islands, New Zealand
Invades wet forests	Now invading in South Pacific islands, New Zealand, Caribbean islands, South Africa
Displaces native plants in forests and along streams; sap can cause serious chemical burns	Now invading in the United States, where it is a federally listed noxious weed; also in Great Britain, central and northern Europe
Dominates in moist meadows and forests; included in wildflower seed mixes	Avoid near wildlands; now invading in much of the United States
Invades along streams, in wetlands and wet forests	Now invading in parts of the United States, Great Britain, most of Europe

Species	Common Name(s)	Native To
HERBACEOUS PLANTS *(continued)*		
***Lamiastrum galeobdolon* 'Florentinum'**	Yellow archangel, golden dead-nettle	Temperate Asia
Lavandula stoechas	Spanish lavender	Southern Europe
Leonotis nepetifolia	Lion's ear	Tropical Africa
Leucanthemum vulgare*, syn. *Chrysanthemum vulgare	Ox-eye daisy	Europe
Linaria dalmatica*, syn. *Linaria genistifolia* ssp. *dalmatica	Dalmatian toadflax	Europe
Linaria vulgaris	Yellow toadflax, butter and eggs	Europe
Lunaria annua	Moneyplant	Southeastern Europe
Mikania micranatha	Mile-a-minute weed	Tropical America
Myosotis scorpioides	Forget-me-not	Eurasia
Myosotis sylvatica	Wood forget-me-not	Eurasia
Oxalis pes-caprae	Soursop, Bermuda buttercup	South Africa
Ruellia brittoniana	Mexican petunia, Britton's wild petunia	Mexico
Senecio elegans	Purple groundsel, purple ragwort	South Africa
Sphagneticola trilobata*, syn. *Wedelia trilobata	Wedelia	Tropical America

Impacts	Cautions for Use
Fast-growing ground cover dominates quickly	Avoid in temperate areas; now invading in the western United States
Spreads into some wildlands, but impacts are unknown	Now invading in California, temperate South America, Australia, New Zealand, South Africa
Can form dense thickets along streams and rivers	Avoid near natural areas; now invading in Hawaii, southeastern United States, Australia
Found in grasslands, wet meadows; carries viruses that can spread to crops; often included in wildflower seed mixes	Do not use anywhere; now invading in forty countries on several continents
Uses extensive creeping roots to outcompete native species	Do not use anywhere; now invading in North America, Russia
Spreads aggressively with seeds and creeping roots	Do not use anywhere; now invading in North America, South America, Russia, New Zealand, Australia, South Africa, and elsewhere
Spreads in open and shady forests; can be controlled by removing the fruits before they mature	Avoid in temperate areas
Fast-growing vine can smother native plants	Now invading in the United States, South Pacific, tropical Asia
Aggressive in moist areas along streams	Avoid near wildlands; now invading in the United States, Australia
Spreads in upland forests; often found in wildflower seed mixes	Avoid near wildlands
Found in dunes and dry areas	Avoid in Mediterranean climates
Persistent	Avoid near wildlands
Sows freely, but impacts are unclear	Now invading in California, New Zealand, Australia; monitor everywhere
Grows rapidly; can form a dense cover, which prevents the regeneration of native species; spreads vegetatively	Avoid near natural areas; now invading in humid areas of the United States, South Pacific islands

Species	Common Name(s)	Native To
HERBACEOUS PLANTS *(continued)*		
Tradescantia fluminensis*, syn. *T. albiflora	White-flowered wandering Jew	Tropical South America
Tradescantia spathacea*, syn. *Rhoeo spathacea	Oyster plant, Moses in a boat	Tropical America
Tradescantia zebrine*, syn. *Zebrina pendula	Wandering Jew	Mexico
Verbena bonariensis	Purple top	South America
Veronica persica	Persian speedwell	Eurasia
Vinca major	Greater periwinkle	Europe, North Africa
Zantedeschia aethiopica	Calla lily, arum lily	South Africa
AQUATIC PLANTS		
Butomus umbellatus	Flowering rush	Eurasia
Cryptocoryne beckettii	Watertrumpet	Sri Lanka
Egeria densa	Brazilian elodea	Neotropics
Eichhornia crassipes	Water hyacinth	Brazil
Elodea canadensis	Elodea, Canadian pondweed	North America
Hydrilla verticillata	Hydrilla	India, Korea
Hydrocotyle ranunculoides	Hydrocotyl, floating pennywort	North America

Impacts	Cautions for Use
Forms dense ground cover with mats up to sixty centimeters thick; smothers native species; prevents regeneration in moist forests	Avoid in humid areas; now invading in southeastern United States, New Zealand, Australia (New South Wales)
Invades dry shaded areas	Now invading in southeastern United States, South Pacific
Forms dense carpets that smother native plants	Now invading in Australia (Queensland), South Pacific islands
Seeds heavily; forms dense stands that exclude native species	Avoid in grasslands; now invading in southeastern United States, southeastern Australia, South Africa
Forms dense carpets that smother native plants	Now invading in many parts of the United States, Europe, Australia
Covers native plants, especially along streams and rivers	Avoid in warm Mediterranean climates
Forms dense fields in moist, warm areas	Now invading in parts of Australia
Outcompetes shoreline vegetation	Now invading in eastern and midwestern United States, southern Canada
Forms dense stands in rivers	Avoid near freshwater in warm areas; now invading in Florida, Texas
Forms dense mats that outcompete native plants and impede boat traffic	Now invading in much of the United States, Chile, England, New Zealand, Australia
Forms dense mats that outcompete native plants and impede boat traffic	Avoid in warm climates; now invading in tropical and warm temperate areas worldwide
Forms dense mats that outcompete native plants and impede boat traffic	Avoid near temperate ponds and lakes; now invading in Great Britain, northern Europe, much of Australia
Forms dense mats that outcompete native plants and impede boat traffic	Now invading in much of the temperate world
Forms dense mats that outcompete native plants and impede boat traffic	Now invading in the Neotropics, Europe, western Australia

Species	Common Name(s)	Native To
AQUATIC PLANTS *(continued)*		
Hygrophila polysperma	Green hygro, Indian swamp weed	India
Lysimachia vulgaris	Yellow garden loosestrife	Eurasia
Lythrum salicaria	Purple loosestrife	Europe
Myriophyllum aquaticum	Parrot's feather	Brazil
Myriophyllum spicatum	Milfoil	Eurasia
Nymphaea odorata	Fragrant water lily	Eastern North America
Pistia stratiotes	Water lettuce	Unknown
Salvinia molesta	Giant salvinia, water fern	Brazil
CLIMBERS		
Abrus precatorius	Rosary pea	Indonesia
Ampelopsis brevipedunculata	Porcelain berry	Northeastern Asia
Anredera cordifolia	Madeira vine	Tropical America
Araujia sericifera	Moth plant	Peru
Cardiospermum grandiflorum	Balloon vine	Neotropics
Cayratia japonica	Bushkiller, sorrel vine	Tropical Asia

Impacts	Cautions for Use
Forms dense mats that outcompete native plants and impede boat traffic; may be able to invade cool temperate climates	Avoid in warmer climates; now invading in fast-moving streams in southeastern United States
Forms dense stands along the edge of lakes and in wetlands that exclude native plants and prevent waterfowl from coming onto land	Avoid in temperate areas; now invading in many parts of the United States
Forms dense stands along the edge of lakes and in wetlands; has wide climate tolerance	Do not use anywhere outside of native range
Forms dense mats that outcompete native plants	Avoid near water systems
Forms dense mats that outcompete native plants	Avoid near water systems
Forms dense mats that outcompete native plants and impede boat traffic	Avoid in temperate areas; now invading in western United States
Forms dense mats that outcompete native plants and impede boat traffic	Do not use anywhere; now invading in at least forty countries
Forms dense mats that outcompete native plants and impede boat traffic; grows fast	Do not use anywhere
Woody vine climbs into trees; hard to control	Do not use anywhere; now invading in tropics
Covers and smothers trees and shrubs	Avoid in temperate areas; now invading in eastern United States
Covers and smothers small trees and shrubs	Avoid in tropical areas; now invading in New Zealand, Australia, South Africa, several South Pacific islands
Covers and smothers small trees and shrubs	Do not use anywhere; now invading in western United States, New Zealand, South Africa, eastern Australia, Israel
Uses tendrils to climb trees; smothers native plants	Avoid in wet tropics; now invading in Hawaii, Australia, Cook Islands
Aggressive vine smothers trees	Now in southeastern United States; monitor everywhere

Species	Common Name(s)	Native To
CLIMBERS *(continued)*		
Celastrus orbiculatus	Oriental bittersweet	Asia
Clematis vitalba	Traveler's joy, old man's beard	Europe
Coccinia grandis	Ivy gourd, scarlet gourd	Africa, Asia
Cryptostegia grandiflora	Rubber vine	Madagascar
Delairea odorata,* syn. *Senecio mikanioides	Cape ivy, German ivy	South Africa
Dioscorea bulbifera	Air potato	Tropical Asia
Epipremnum pinnatum* 'Aureum', syn. *Scindapsus aureus	Golden pothos	Solomon Islands
Ficus pumila	Creeping fig	Asia
***Hedera helix, H. hibernica* (closely related and nearly indistinguishable species)**	English ivy	Europe, Great Britain
Ipomoea indica	Blue morning glory	Mexico, Central America
Jasminum dichotomum	Gold Coast jasmine	Western Africa
Jasminum fluminense	Brazilian jasmine	Africa
Lathyrus latifolius	Perennial sweet pea, everlasting pea	Europe
Lathyrus tingitanus	Tangier pea	Western Mediterranean, Azores

Impacts	Cautions for Use
Woody vine smothers trees and shrubs; increases storm damage by adding the weight from its woody stems	Avoid in temperate areas; now invading in northeastern United States
Aggressive vine smothers even tall trees; hard to control	Avoid in temperate areas; now invading in northwestern United States, New Zealand
Covers and smothers small trees and shrubs	Avoid in warm, humid climates; now invading in southeastern United States, South Pacific islands, Caribbean, Australia
Covers and smothers small trees and shrubs	Do not use anywhere; now invading in South Pacific, Caribbean islands, Australia (Queensland)
Covers and smothers small trees and shrubs; poisonous to mammals and fish	Avoid in Mediterranean climates; now invading in California, Australia, Italy
Covers and smothers small trees and shrubs; can spread during control work by means of small tubers on stems	Now invading in Florida, Africa
Covers and smothers small trees and shrubs; mostly spreads vegetatively	Avoid near wildlands; now invading in Florida, South Pacific islands
Slower growing than some vines, but can create a smothering mass	Now invading in places where a specialist pollinator is present in Hawaii, Texas, Florida; use with caution in other tropical areas
Covers the ground; outcompetes native species; increases storm damage in trees	Now invading in eastern and western United States, Hawaii, Australia, New Zealand, Brazil; some cultivars also invade
Covers and smothers small trees and shrubs	Avoid near wildlands in the tropics; now invading in Australia, South Africa
Covers and smothers small trees and shrubs	Avoid in tropical areas; now invading in Florida, Hawaii
Covers and smothers trees and shrubs	Avoid in tropical areas; now invading in Florida, Hawaii, Caribbean islands, Guam
Covers and smothers shrubs; can be difficult to control	In North America and other temperate climates, grow with caution
Vines are believed to create a fire hazard in western Australia	Now invading in western United States, western and southern Australia, New Zealand

Species	Common Name(s)	Native To
CLIMBERS *(continued)*		
Lygodium japonicum	Japanese climbing fern	Japan
Lygodium microphyllum	Old World climbing fern	Africa, Asia, Australia
Macfadyena unguis-cati	Cat's claw creeper	Neotropics
Merremia tuberosa	Wood rose	Neotropics
Passiflora suberosa	Corky passion vine	Galápagos Islands
Passiflora tarminiana,* syn. *P. mollisima	Banana poka, banana passion fruit	Tropical South America
Pueraria montana* var. *lobata	Kudzu	Asia
Wisteria floribunda	Japanese wisteria	Japan
Wisteria sinensis	Chinese wisteria	China
GRASSES, BAMBOOS, SEDGES, RUSHES, AND RESTIOS		
Ammenophila arenaria	European beach grass	Europe
Arundo donax	Giant reedgrass	India
Cortaderia jubata	Jubata grass	South America
Cortaderia selloana	Pampas grass	South America

Impacts	Cautions for Use
Forms dense, intertwined mats that smother trees and shrubs	Avoid in dry and wet forests; now invading in southeastern United States
Forms, dense intertwined mats that smother trees and shrubs; may increase fire intensity	Avoid in warm climates; now invading in Florida
Forms dense mats on the forest floor and shrubs that exclude native species; underground tubers increase its invasive spread	Avoid in tropical climates; now invading in southeastern United States, Hawaii, eastern Australia, New Caledonia
Covers, smothers, and sometimes pulls down trees and shrubs	Avoid near natural areas in the tropics
Covers and smothers trees and shrubs	Avoid in dry areas; now invading in many South Pacific and Caribbean islands, Australia, Réunion, South Africa
Covers and smothers trees and shrubs	Avoid in wet tropical areas; now invading in Hawaii, New Zealand, South Africa, Asia
Aggressive vine that quickly covers, smothers, and breaks even large trees	Do not use anywhere; now invading in southeastern United States, several South Pacific islands; spreading into cooler areas
Woody vines can smother trees and increase storm damage	Now invading in southeastern United States; should be monitored near wildlands; hybrids with *W. sinensis* may also be invading
Impacts as for *W. floribunda*	Now invading in southeastern United States; should be monitored near wildlands; hybrids with *W. floribunda* may also be invading
Widely planted to stabilize dunes, but reduces native plant and arthropod species	Avoid in temperate areas
Outcompetes native plants; does not provide food or habitat for native animals; transforms floodplains and rivers by fragmenting during floods to form new populations	Do not use anywhere; now invading in western United States, Mexico, Caribbean and South Pacific islands, Australia, South Africa, southeastern Asia
Dominates grasslands; displaces native species	Now invading in California; should be closely monitored everywhere
Invades grasslands	Do not use anywhere in California; studies show that even "sterile" cultivars can produce seeds

Species	Common Name(s)	Native To
GRASSES, BAMBOOS, SEDGES, RUSHES, AND RESTIOS *(continued)*		
Imperata cylindrica	Cogongrass	Southeast Asia
Miscanthus sinensis	Chinese silver grass	Asia
Pennisetum setaceum	Fountain grass	Africa, Middle East
Phragmites australis	Common reed	Europe and the United States (European genotypes are invading in the United States)
Phyllostachys aurea	Golden bamboo	Asia
BULBOUS PLANTS		
Chasmanthe floribunda	African cornflag	South Africa
Crocosmia × crocosmiflora	Montbretia	South Africa
Gladiolus caryophyllaceus	Wild gladiolus	South Africa
Iris pseudacorus	Yellow flag	Europe
Lilium formosanum	Taiwan lily	Taiwan
Ornithogalum umbellatum	Star of Bethlehem	Europe, North Africa
Sparaxis bulbifera, **syn. *Ixia bulbifera***	Harlequin flower	South Africa
Watsonia meriana **var. *bulbillifera***	Bulbil watsonia	South Africa

Impacts	Cautions for Use
Infests pine woodlands, dunes, wetlands, and grasslands; outcompetes native species; increases fire danger	The green wild type is now invading in seventy-three countries and should never be grown; red-tipped cultivars are less or not invasive
Found along roadsides, at forest edges, and in clearings; may form large populations	Avoid in temperate areas; now invading in eastern United States
Can outcompete native species; dry material creates a fire hazard	Avoid in warm areas; now invading in Hawaii, North America, Fiji, South Africa, Australia
Extremely invasive in wetlands and along lakes and other water bodies; forms dense rhizomatous stands	Do not use anywhere outside of native range
Aggressively grows rhizomes; can penetrate root barriers in gardens	Avoid near wildlands; now invading in southern United States, Hawaii, Australia (Queensland)
Colonizes open areas; may exclude native species	Now invading in western United States, southern Australia, New Zealand; should be monitored in Mediterranean climates
Expands vigorously when the corms attached to its rhizomes are moved by soil disturbances; locally abundant but impacts are unclear	Avoid in California, Australia, areas with mild temperatures in Europe
Spreads rapidly in open areas	Now invading in southwestern Australia; should be monitored in Mediterranean climates
Forms dense rhizomatous stands that exclude native shoreline vegetation and make it difficult for waterfowl to get from water to land	Avoid near natural waterways; now invading in many parts of the world
Reproduces by both seeds and bulblets, becoming abundant	Now invading in eastern Australia; should be monitored wherever grown
Invades disturbed areas and some open woods; may impact native vegetation	Now invading in eastern North America; use with caution
Invades wetlands	Now invading in southern and western Australia; use with caution
Grows along creeks and in open areas; can form dense colonies that exclude native species; spreads slowly but difficult to control	Now invading in several parts of Australia; use with caution

Species	Common Name(s)	Native To
CACTI AND SUCCULENTS		
Bryophyllum delagoense*, syn. *Kalanchoe delagoense	Mother of millions, chandelier plant	Madagascar
Carpobrotus edulis	Hottentot fig	South Africa
Mesembryanthemum crystallinum	Crystalline iceplant	South Africa
Opuntia ficus-indica	Common prickly pear	Probably Mexico
FERNS		
Angiopteris evecta	Mule's foot fern	Australia
Cyathea cooperi	Australian treefern	Eastern Australia
Nephrolepis cordifolia	Tuberous sword fern	Australia
Nephrolepis multiflora	Asian sword fern	Tropical Asia

Impacts	Cautions for Use
Invades dry forests and open areas	Now invading in Hawaii, the Virgin Islands, Australia (Queensland), South Africa
Grows into a thick mat that suppresses native vegetation	Avoid in Mediterranean climates; now invading in coastal California, the Mediterranean, New Zealand
Outcompetes native plants for water; can accumulate unusually high levels of nitrate salts under it, which prevents the establishment of native species	Avoid in Mediterranean climates; now invading in California and Arizona
Forms impenetrable thickets	Avoid near wildlands; now invading in parts of Africa, nearly all of Australia, the Caribbean islands, the Mediterranean countries
Forms dense stands that displace and shade out native species	Avoid near moist tropical forests; now invading in Hawaii, Costa Rica, and Jamaica
Can dominate high-quality forests	Avoid in warm, wet climates; now invading in Hawaii, western Australia, French Polynesia
Colonizes shady forests; can spread aggressively	Avoid in warm, moist climates; now invading in southeastern United States
Aggressively spreads along roadsides, trails, and hardened lava flows	Now invading in Hawaii, Puerto Rico, Caribbean islands

GLOSSARY

Allelopathy The ability of some plants to release chemicals that may inhibit growth in other plants.

Annual A plant that completes its entire life cycle, from seed germination through flower and fruiting to death, within one year. Some annuals germinate in the fall and overwinter to flower the next year, while most germinate and flower in a single growing season. Compare *perennial*.

Cultivar Short for *cultivated variety*. Cultivars generally arise through hybridizing or careful selection of characteristics such as growth form, flower or leaf color, or drought tolerance. They are usually maintained through asexual propagation such as cuttings but may also reproduce sexually, through seeds. In botanical naming they are differentiated from naturally occurring varieties by the use of an epithet in single quotes, e.g., *Arctostaphylos uva-ursi* 'Massachusetts'.

Deciduous Falling off or having parts that fall off, such as leaves or flower parts. Usually refers to trees that shed all their leaves in autumn. Compare *evergreen*.

Ecosystem An environment with interdependent biological, physical, and chemical components.

Ecosystem services The benefits that ecosystems provide humans, such as the cleansing of water through soil filtration or of air by the carbon uptake of trees.

Evergreen Having leaves throughout the year. No leaf lasts forever, but because there are always leaves on the plant, it is always (ever) green. Most conifers are evergreen, as are many flowering plants. Compare *deciduous.*

Fitness The sum of genetic advantages an organism has in a given environment that allow it to survive and reproduce. The fittest organisms produce the largest number of offspring that reach reproductive maturity and pass their genes on to future generations.

Genotype The genetic composition of an organism. Populations of species that reproduce asexually may have only one genotype, but those that reproduce sexually may have many.

Greenhouse gas A gas in the lower atmosphere that absorbs solar heat reradiated by Earth's surface. Greenhouse gases occur naturally, but human activities produce more of certain key ones, including carbon dioxide (CO_2), methane (CH_4), and nitrous oxide (NO_2).

Herbaceous Having no woody aboveground parts, such as trunks, stems, or branches. Annual bedding plants like marigolds and petunias are herbaceous.

Humus Organic material in the soil that has decomposed as far as possible under current conditions.

Impervious surface Any surface that blocks water from draining into the soil. Examples include concrete sidewalks, asphalt driveways, mortared stone patios, and most urban rooftops.

Microorganism A microscopic organism, usually a bacterium, fungus, protozoan, or virus, such as those found in soil.

Mineral Composed of usually crystalline elements or chemical compounds formed by geological processes. Mineral soils have a high concentration of sand, clay, or similar material and a low organic content. Compare *organic*.

Nitrogen fixation The conversion of atmospheric nitrogen into forms that plants can use, such as ammonia and nitrates. Some plants—many in the legume family but also in such genera as *Alnus, Myrica,* and *Casuarina*—use nodules on their roots to host bacteria that facilitate this.

Organic (1) Containing carbon or, more specifically, carbon-based organisms or their decomposing materials (e.g., organic soil, which has a high amount of decaying plant and animal parts). Compare *mineral*. (2) Grown or prepared without chemically formulated substances, such as pesticides or fertilizers. Commercial goods must be produced under rules adopted by the United States Department of Agriculture to be labeled organic.

Pathogen An agent that causes disease, including bacteria, fungi, and viruses.

Perennial A species capable of surviving more than three years under normal growing conditions and of producing flowers and fruits annually. Compare *annual*.

pH A measure of acidity or alkalinity, based on the concentration of hydrogen ions in an aqueous solution and denoted on a logarithmic scale. If the pH is less than 7 (the midpoint or neutral number), the substance is acidic and has a high concentration of hydrogen ions. If the number is higher than 7, the substance is alkaline (or basic) and does not have many hydrogen ions. Because the scale is logarithmic, the differences in numbers can be deceptive: for instance, a pH of 6 means ten times more hydrogen ions than a pH of 7 and one hundred times more than a pH of 8.

Population A group of individuals within a single species that are close enough—depending on the species and its reproductive system—to interbreed.

Rhizome A horizontal stem, often underground, that bears roots, shoots, and leaves and is capable of persisting from year to year.

Species Often defined as a group of organisms that closely resemble one another (the word derives from the Latin *specere*, "to look") and can interbreed within the group but not with other relatives. Classifying plants this way is often difficult because when barriers to gene movement are removed or circumvented, as through plant introduction, they may be able to hybridize with close relatives with which they now co-occur.

Stomata Openings (singular: *stoma* or *stomate*) usually on the underside of leaves that allow carbon dioxide in for photosynthesis and also transpire water.

Sustainable Capable of maintaining functions and productivity into the future, as in use of materials that does not compromise future generations' access to the same materials to meet their own needs. The core idea is that natural resources must be used in a way and at a rate such that they can be adequately replenished.

Transpiration The passage of water vapor from a plant into the surrounding atmosphere, usually through stomata.

Vascular plants Plants—including ferns, conifers, and flowering plants—that have tissues (xylem and phloem) through which water and nutrients move.

Vector A pathway by which plants move. This can include businesses, such as nurseries that import plants from different states, or animals that move seeds that get caught in their fur or that they ingest and later defecate.

Vegetative reproduction Asexual plant reproduction through such means as rhizomes or the ability of stems to form roots when they touch the ground for a period of time (layering). Some species, including blackberries and some grasses, can produce unfertilized seeds asexually, and this method of reproduction is also sometimes considered vegetative.

RESOURCES

Introduction

Leopold, A. 1949. *A Sand County Almanac, and Sketches Here and There.* Oxford University Press, New York.

1. The Skin of the Earth

References for Gardeners

Baskin, Y. 2005. *Under Ground: How Creatures of Mud and Dirt Shape Our World.* Island Press, Washington, DC.

Campbell, S. 1998. *Let It Rot! The Gardener's Guide to Composting.* Storey Press, Pownal, VT.

Cornell University. http://compost.css.cornell.edu/Composting_homepage.html. This site has several fact sheets about composting.

Dunn, N., ed. 2009. *Healthy Soils for Sustainable Gardens.* Brooklyn Botanic Garden All-Region Guides.

Environment and Human Health. www.ehhi.org/reports/turf/. The report on recycled rubber tire mulch can be downloaded here.

Martin, D., and G. Gershuny, eds. 1992. *The Rodale Book of Composting.* Rodale Press, Emmaus, PA.

Ohio State University. http://ohioline.osu.edu/com-fact/0001.html. This page has good information about composting.

Organic Gardening. www.organicgardening.com/feature/0,7518,s1-3-81-185,00.html. You can find some recipes for soil mixes here—just remember to substitute one of the peat alternatives.

Plantlife International. www.plantlife.org.uk/campaigns/plants_peat. Provides information about peat mining and efforts to stop horticultural peat moss use.

Ramsar Convention on Wetlands. www.ramsar.org.

Sachs, P.D. 1993. *Edaphos: Dynamics of a Natural Soil System*. The Edaphic Press, Newbury, VT.

Sammis, T. "Soil Texture Analysis." http://weather.nmsu.edu/teaching_Material/soil456/soiltexture/soiltext.htm.

Singer, M.J., and D.N. Munns. 2002. *Soils: An Introduction*. 5th ed. Prentice Hall, Upper Saddle River, NJ.

University of Massachusetts Amherst. www.umass.edu/plsoils/soiltest/. For a small fee the Department of Plant and Soil Sciences will test your soil.

University of Minnesota Duluth. www.nrri.umn.edu/worms/default.htm. This is an instructive webpage about earthworms.

Technical Papers

Bever, J.D., P.A. Schultz, A. Pringle, and J.B. Morton. 2001. Arbuscular Mycorrhizal Fungi: More Diverse than Meets the Eye, and the Ecological Tale of Why. *BioScience* 51(11): 923–931.

Bohlen, P.J., S. Scheu, C.M. Hale, M.A. McLean, S. Migge, P.M. Groffman, and D. Parkinson. 2004. Non-native Invasive Earthworms as Agents of Change in Northern Temperate Forests. *Frontiers in Ecology and the Environment* 2(8): 427–435.

Chalker-Scott, L. 2007. Impact of Mulches on Landscape Plants and the Environment—A Review. *Journal of Environmental Horticulture* 25: 239–249.

Greenly, K., and D. Rakow. 1995. The Effects of Mulch Type and Depth on Weed and Tree Growth. *Journal of Arboriculture* 21: 225–232.

Hale, C.M., and G.E. Host. 2005. Assessing the Impacts of European Earthworm Invasions in Beechmaple Hardwood and Aspen-Fir Boreal Forests of the Western Great Lakes Region. National Park Service Great Lakes Inventory and Monitoring Network Report GLKN/2005/11.

Jonsson-Ninniss, S., and J. Middleton. 1991. Effect of Peat Extraction on the Vegetation in Wainfleet Bog, Ontario. *Canadian Field-Naturalist* 105: 505–511.

Linderman, R.G., and E.A. Davis. 2004. Evaluation of Commercial Inorganic and Organic Fertilizer Effects on Arbuscular Mycorrhizae Formed by *Glomus intraradices*. *HortTechnology* 14(2): 196–202.

Long, C.E., B.L. Thorne, N.L. Breisch, and L.W. Douglass. 2001. Effect of Organic and Inorganic Landscape Mulches on Subterranean Termite (Isoptera: Rhinotermitidae) Foraging Activity. *Environmental Entomology* 30: 832–836.

Lowry, C. A., et al. 2007. Identification of an Immune-Responsive Mesolimbocortical Serotonergic System: Potential Role in Regulation of Emotional Behavior. *Neuroscience* 146 (2–5): 756–772.

Pickering, J. S., and A. Shepherd. 2000. Evaluation of Organic Landscape Mulches: Composition and Nutrient Release Characteristics. *Arboriculture Journal* 23: 175–187.

Redman, R. S., K. B. Sheehan, R. G. Stout, R. J. Rodriguez, and J. M. Henson. 2002. Thermotolerance Generated by Plant/Fungal Symbiosis. *Science* 298: 1581.

Rochefort, L., F. Quinty, S. Campeau, K. Johnson, and T. Malterer. 2003. North American Approach to the Restoration of *Sphagnum*-Dominated Peatlands. *Wetlands Ecology and Management* 11: 3–20.

Steward, L. G., T. D. Sydnor, and B. Bishop. 2003. The Ease Of Ignition of 13 Landscape Mulches. *Journal of Arboriculture* 29: 317–321.

White, T. C. R. 1984. The Abundance of Invertebrate Herbivores in Relation to the Availability of Nitrogen in Stressed Food Plants. *Oecologia* 62: 90-105.

2. Water, Our Most Precious Resource

References for Gardeners

Bannerman, R. 2003. *Rain Gardens: A How-To Manual for Homeowners.* University of Wisconsin Extension Publications, Madison, WI.

Bormann, F. H., D. Balmori, and G. Geballe. 1993. *Redesigning the American Lawn: A Search for Environmental Harmony.* Yale University Press, New Haven, CT.

Brooklyn Botanic Garden. 1995. *The Natural Lawn and Alternatives.* Brooklyn Botanic Garden Handbook #136.

Dunnett, N. 2004. *Planting Green Roofs and Living Walls.* Timber Press, Portland, OR.

Dunnett, N., and A. Clayden. 2007. *Rain Gardens: Managing Water Sustainably in the Garden and Designed Landscape.* Timber Press, Portland, OR.

Ellefson, C. 1992. *Xeriscape Gardening: Water Conservation for the American Landscape.* Macmillan, New York, NY.

Glennon, R. 2009. *Unquenchable: America's Water Crisis and What to Do about It.* Island Press, Washington, DC.

Hyland, B. 1992. "Xeriscaping Demystified," in *The Environmental Gardener,* pages 21–25. Brooklyn Botanic Garden Handbook #130.

International Rain Catchment Systems Association. www.eng.warwick.ac.uk/ircsa. This is a good website with lots of links and fact sheets.

Melchert, D. B. 1992. "How to Create a Streamside Garden, Including the Stream," in *The Environmental Gardener,* pages 45–55. Brooklyn Botanic Garden Handbook #130.

Michigan State University. www.hrt.msu.edu/greenroof/. This home page of the MSU Green Roof Research Program has background and new information on green roofs.

Royal Horticulture Society. www.rhs.org.uk/advice/profiles1105/drought.asp. The society's page on drought-resistant gardening also has links to more subjects.

Taylor, J. 1993. *Dought-Tolerant Plants: Waterwise Gardening for Every Climate.* Prentice Hall, New York, NY.

Technical Papers

Bortleson, G. C., and D. A. Davis. 1997. Pesticides in Selected Small Streams in the Puget Sound Basin, 1987–1995. U.S. Geological Survey Fact Sheet 067-97.

Gackstatter, J. A., A. F. Bartsch, and C. A. Callahan. 1978. The Impact of Broadly Applied Effluent Phosphorus Standards on Eutrophication Control. *Water Resources Research* 14: 1155–1158.

Gilliom, R. J. 2007. Pesticides in U.S. Streams and Groundwater. *Environmental Science and Technology* 41: 3407–3413.

Laetz, C. A., D. H. Baldwin, T. K. Collier, V. Hebert, J. D. Stark, and N. L. Scholz. 2009. The Synergistic Toxicity of Pesticide Mixtures: Implications for Risk Assessment and the Conservation of Endangered Pacific Salmon. *Environmental Health Perspectives,* 117: 348–353.

Paul, M. J., and J. L. Meyer. 2001. Streams in the Urban Environment. *Annual Review of Ecology and Systematics* 32: 333–365.

Robbins, P., A. Polderman, and T. Birkenholtz. 2001. An Ecology of the City. *Cities* 18: 369–380.

Tierney, K., J. L. Sampson, P. S. Ross, M. A. Sekela, and C. J. Kennedy. 2008. Salmon Olfaction Is Impaired by an Environmentally Realistic Pesticide Mixture. *Environmental Science and Technology* 42: 4996-5001.

3. Should You Go Native?

There are many books with recommendations for native plants appropriate to specific regions. Check with your local bookstores and libraries.

References for Gardeners

Center for Plant Conservation. www.centerforplantconservation.org. Check this website for information about rare native plant conservation and conservation gardens near you.

Convention on International Trade in Endangered Species. www.cites.org. Contains information about the CITES treaty.

Environmental Protection Agency. www.epa.gov/heatisland/pilot/index.htm. This site gives links to more data from the Urban Heat Island Pilot Project.

International Union for the Conservation of Nature. www.iucnredlist.org. The IUCN regularly analyzes species and publishes lists of those considered globally rare.

Johnson, L. 1998. *Grow Wild! Low-Maintenance, Sure Success, Distinctive Gardening with Native Plants.* Fulcrum Publishing, Golden, CO.

Ladybird Johnson Wildflower Center. www.wildflower.org/collections. This website supplies a list of appropriate native plants for each state, along with photographs and instructions for growing each species.

Marinelli, J., ed. 1996. *Going Native: Biodiversity in Our Own Backyards.* Brooklyn Botanic Garden Handbook #140.

National Aeronautics and Space Administration. "Urban Heat Islands Make Cities Greener." www.nasa.gov/centers/goddard/news/topstory/2004/0801uhigreen.html. This article has details about urban heat islands.

Ottesen, C. 1995. *The Native Plant Primer.* Harmony Books, New York.

Plant Conservation Alliance. www.nps.gov/plants. Bureau of Land Management. The definition of *native* is at www.nps.gov/plants/restore/pubs/intronatplant/whyusenatives.htm.

Plant Natives. www.plantnative.org. Although it perpetuates some of the misconceptions discussed in this chapter, this organization provides step-by-step instructions to planning a native plant garden and resources to help locate nurseries selling native plants in your area.

U.S. Department of Agriculture Plant Database. http://plants.usda.gov. The Department of Agriculture maintains a searchable database of native and invasive plants.

U.S. Forest Service. www.fs.fed.us/wildflowers/nativeplantmaterials/careaboutgenetics.shtml. This excellent site goes into more depth about the genetic issues of growing native plants.

Wasowski, S., and A. Wasowski. 2000. *The Landscaping Revolution: Garden with Mother Nature, Not against Her.* Contemporary Books, Lincolnwood, IL.

Technical Papers

Cook, S. A. 1962. Genetic System, Variation, and Adaptation in *Eschscholzia californica. Evolution* 16: 278–299.

Ellstrand, N. C., H. C. Prentice, and J. F. Hancock. 1999. Gene Flow and Introgression from Domesticated Plants into Their Wild Relatives. *Annual Review of Ecology and Systematics* 30: 539–563.

Hummel, R. L., and R. Maleike. 1998. Water Use of Native and Introduced Plants. *B&B* 50: 1, 6, 12.

Levitt, D. G., J. R. Simpson, and J. L. Tipton. 1995. Water Use of Two Landscape Tree Species in Tucson, Arizona. *Journal of the American Society for Horticultural Science* 120: 409–416.

Lo, C. P., and D. A. Quattrochi. 2003. Land-Use and Land-Cover Change, Urban Heat Island Phenomenon, and Health Implications: A Remote Sensing Approach. *Photogrammetric Engineering and Remote Sensing* 69: 1053–1063.

Montalvo, A. M., and N. C. Ellstrand. 2001. Nonlocal Transplantation and Outbreeding Depression in the Subshrub *Lotus scoparius* (Fabaceae). *American Journal of Botany* 88: 258–269.

Potts, B. M., R. C. Barbour, A. B. Hingston, and R. E. Vaillancourt. 2003. Genetic Pollution of Native Eucalypt Gene Pools—Identifying the Risks. *Australian Journal of Botany* 51: 1–25.

Schierenbeck, K. A., R. N. Mack, and R. R. Sharitz. 1994. Effects of Herbivory on Growth and Biomass Allocation in Native and Introduced Species of *Lonicera. Ecology* 75: 1661–1672.

Schmidt, K. A., and C. J. Whelan. 1999. Effects of Exotic *Lonicera* and *Rhamnus* on Songbird Nest Predation. *Conservation Biology* 13: 1502–1506.

4. Aliens among Us

References for Gardeners

Baskin, Y. 2002. *A Plague of Rats and Rubbervines: The Growing Threat of Species Invasions.* Island Press, Washington, DC. This is a useful background book written for the general public.

Center for Invasive Species and Ecosystem Health, University of Georgia. www.invasive.org. This website has information about invasive plants and plant pests and hosts the archives of the Nature Conservancy's Global Invasive Species Team (disbanded in March 2009), which include Stewardship Abstracts, photographs, humorous videos about weedy species, and much more.

Center for Plant Conservation. www.centerforplantconservation.org/invasives/codesN.html. This page has links to the St. Louis workshop's findings and principles and all the Codes of Conduct.

Marinelli, J., and J. Randall. 1996. *Weeds of the Global Garden*. Brooklyn Botanic Garden Handbook Series.

Myers, J., and D. Bazely. 2003. *Ecology and Control of Introduced Plants*. Cambridge University Press, Cambridge, UK. This book is more technical, but it also provides good background information.

National Association of Exotic Pest Plant Councils. www.naeppc.org. This organization coordinates with local groups to address invasive issues. The site will also help you find the nearest Pest Plant Council.

National Invasive Species Information Center. www.invasivespeciesinfo.gov. A comprehensive website from the USDA's Agricultural Library, with notices about regulation changes and new research.

National Park Service. www.nps.gov/plants/alien. Weeds Gone Wild: Alien Plant Invaders of Natural Areas is the web-based project of the Plant Conservation Alliance's Alien Plant Working Group.

PlantRight. www.plantright.org. This page provides guidelines for the horticultural community designed in conjunction with the California Horticultural Invasives Prevention partnership. The "Regional invasive and alternative plants" link will help you find noninvasive alternatives.

Weber, E. 2003. *Invasive Plant Species of the World*. CABI Publishing, Cambridge, MA. This book provides information on 450 species invading around the world.

Technical Papers

Crooks, J. A., and M. E. Soulé. 1999. Lag Times in Population Explosions of Invasive Species. In O. Sandlund, P. J. Schei, and A. Viken, *Invasive Species and Biodiversity Management*. Kluwer Academic Publishers, Dordrecht, Netherlands.

Mühlenbach, V. 1979. Contributions to the Synanthropic (Adventive) Flora of the Railroads in St. Louis, Missouri, U.S.A. *Annals of the Missouri Botanical Garden* 66: 1–108.

Pimentel, D., R. Zuniga, and D. Morrison. 2005. Update on the Environmental and Economic Costs Associated with Alien-Invasive Species in the United States. *Ecological Economics* 52: 273–288.

Reichard, S. 1997. Preventing the Introduction of Invasive Plants. In J. Luken and J. Thieret, eds., *Assessment and Management of Plant Invasions,* pages 215–227. Springer-Verlag, New York.

Urgenson, L., S. Reichard, and C. Halpern. 2009. Community and Ecosystem Consequences of Giant Knotweed *(Polygonum sachalinense)* Invasion into Riparian Forests of Western Washington, USA. *Biological Conservation* 142: 1536–1541.

von der Lippe, M., and I. Kowarik. 2007. Long Distance Dispersal of Plants by Vehicles as a Driver of Plant Invasions. *Conservation Biology* 21: 986–996.

5. The Wild Kingdom

References for Gardeners

Adams, G. 1994. *Birdscaping Your Garden.* Rodale Press, Emmaus, PA.

Arnoux, J. 1996. *The Ultimate Water Garden Book.* Taunton Press, Newtown, CT.

Backyard Birding Page, Baltimore Bird Club. http://baltimorebirdclub.org/by/backyard.html. This page offers links to good basic information on providing habitat for birds.

Brooklyn Botanic Garden. www.bbg.org/gar2/topics/wildlife. This site has links to information on how to attract birds, butterflies, and other creatures to your garden. Some is specific to the northeastern United States, but much is broadly applicable.

Dole, C. H. 2003. *The Butterfly Gardener's Guide.* Brooklyn Botanic Garden Handbook #175.

Fell, D. 1998. *Water Gardening with Derek Fell.* Friedman/Fairfax, New York.

Fish and Wildlife Service. http://wsfrprograms.fws.gov/Subpages/NationalSurvey/reports2006.html. This site has links to reports about wildlife viewing in the United States.

Hadidian, J., G. Hodge, and J. Grandy. 1997. *Wild Neighbors: The Humane Approach to Living with Wildlife.* Fulcrum Publishing, Golden, CO.

Link, R. 2004. *Living with Wildlife in the Pacific Northwest.* University of Washington Press. This book has some specific regional information but also great general advice about both attracting and managing wildlife.

Logsdon, G. 1999. *Wildlife in Your Garden: How to Live in Harmony with Deer, Raccoons, Rabbits, Crows, and Other Pesky Creatures.* Indiana University Press, Bloomington.

Lovejoy, S. 2004. *A Blessing of Toads: A Gardener's Guide to Living with Nature.* Hearst Books, New York. Despite the title, this book covers a wide array of animals.

Marinelli, J. 2008. *The Wildlife Gardener's Guide.* Brooklyn Botanic Garden Handbook #189.

National Wildlife Federation. www.nwf.org/Get-Outside.aspx. This is the portal page to many great garden (and outdoor) resources from the NWF. Go to www.nwf.org/Get-Outside/Outdoor-Activities/Garden-for-Wildlife/Gardening-Tips/Build-a-Bat-House.aspx for clear directions on building a bat house. You can find instructions for creating a toad abode at www.nwf.org/Get-Outside/Be-Out-There/Activities/Observe-and-Explore/Make-a-Toad-House.aspx. For Backyard Wildlife Sanctuary certification, visit www.nwf.org/Get-Outside/Outdoor-Activities/Garden-for-Wildlife.aspx.

Robinson, P. 1997. *The American Horticultural Society Complete Guide to Water Gardening.* DK Publishing, New York.

Urban Bee Gardens. http://nature.berkeley.edu/urbanbeegardens/. This website of the Urban Bee Project of the San Francisco Bay Area will open your eyes to the wonders of bees.

Technical Papers

Beier, P., and R. Noss. 1998. Do Habitat Corridors Provide Connectivity? *Conservation Biology* 12: 1241–1252.

Harper, K.T. 1979. Some Reproductive and Life History Characteristics of Rare Plants and Implications for Management. *Great Basin Naturalist Memoirs* 3: 129–137.

6. Preventing and Managing Pests

Because there are so many potential pest species, it is probably best to take a sample or photograph to an extension agent, garden center, or public garden for assistance. The references below may also be helpful.

References for Gardeners

Best, C. 1992. "Natural Pesticides: Are They Really Safer?" in *The Environmental Gardener,* pages 33–37. Brooklyn Botanic Garden Handbook #130.

Caldwell, B., E. B. Rosen, E. Sideman, A. M. Shelton, and C. Smart. 2005. *Resource Guide for Organic Insect and Disease Management.* www.nysaes.cornell.edu/pp/resourceguide. This guide to organic gardening for farmers can be helpful to home gardeners as well.

Carson, R. 2002 (fortieth anniversary edition). *Silent Spring.* Houghton Mifflin, New York, NY.

Cloyd, R.A., P.L. Nixon, and N.R. Pataky. 2009. *IPM for Gardeners: A Guide to Integrated Pest Management.* Timber Press, Portland, OR.

Cranshaw, W. 2004. *Garden Insects of North America: The Ultimate Guide to Backyard Bugs.* Princeton University Press, Princeton, NJ.

Creasy, R. 1993. *Organic Gardener's Edible Plants.* Van Patten Publishing, Portland, OR.

Daniels, C.H. October 2002. "What's Cooking with Vinegar Recommendations? Acetic Acid as Herbicide." *Agricultural and Environmental News.* http://aenews.wsu.edu/Oct02AENews/Oct02AENews.htm#Vinegar.

Environmental Protection Agency. Pesticide Fact Sheets. www.epa.gov/opp00001/factsheets/ipm.htm. This web page provides a good general outline of the principles of Integrated Pest Management.

Environment, Health and Safety Online. www.ehso.com/msds.php. This site has links to thousands of Material Safety Data Sheets.

Hollingsworth, C., and K. Idoine. 1992. "An Environmental Gardener's Guide to Pest Management," in *The Environmental Gardener,* pages 26–32. Brooklyn Botanic Garden Handbook #130.

MSDS. www.msds.com. This site has a searchable database of millions of Material Safety Data Sheets.

Ogden, D. 1999. *Straight-Ahead Organic: A Step-by-Step Guide to Growing Great Vegetables in a Less Than Perfect World.* Chelsea Green, White River Junction, VT.

Smith, M., and A. Carr. 1988. *Rodale's Garden Insect, Disease, and Weed Identification Guide.* Rodale Press, Emmaus, PA.

UC IPM online. www.ipm.ucdavis.edu. This website hosted by the University of California includes pest management techniques, educational resources, and research projects.

Verdoorn, G. September 2004. "Will the Real Organic Pesticide Please Stand Up?" *Science in Africa.* www.scienceinafrica.co.za/2004/september/organic.htm. This is a no-nonsense article on "organic" pest control.

Technical Papers

Cranshaw, W. 1997. Attractiveness of Beer and Fermentation Products to the Grady Garden Slug. Colorado State University Technical Bulletin TA97-1.

Litterick, A.M., L. Harrier, P. Wallace, C.A. Watson, and M. Wood. 2004. The

Role of Uncomposted Materials, Composts, Manures, and Compost Extracts in Reducing Pest and Disease Incidence and Severity in Sustainable Temperate Agricultural and Horticultural Crop Production—A Review. *Critical Reviews in Plant Sciences* 23: 453–479.

7. Confronting Climate Change

References for Gardeners

Biggs, T. 1980. *Vegetables: The Simon and Schuster Step-by-Step Encyclopedia of Practical Gardening.* Simon and Schuster, New York, NY.

Creasy, R. 1993. *Organic Gardener's Edible Plants.* Van Patten Publishing, Portland, OR. Creasy has written a number of books on food gardening, and all are good; this one focuses on organic methods.

Gore, A. 2006. *An Inconvenient Truth: The Planetary Emergency of Global Warming and What We Can Do About It.* Rodale Books, Emmaus, PA. Both this book and the documentary of the same name provide a good introduction to climate change. The website associated with the film is www.climatecrisis.net.

Greener Choices. www.greenerchoices.org/calculators.cfm. This website, run by Consumers Union, the publisher of *Consumer Reports,* offers links to various calculators, including ones for your carbon footprint.

Intergovernmental Panel on Climate Change. www.ipcc.ch. All the Fourth Assessment Report papers cited in this chapter can be found here, and the Frequently Asked Questions (www.ipcc.ch/pdf/assessment-report/ar4/wg1/ar4-wg1-faqs.pdf) and Summary for Policymakers (www.ipcc.ch/pdf/assessment-report/ar4/wg1/ar4-wg1-spm.pdf) are especially informative.

Ogden, S. 1999. *Straight-Ahead Organic: A Step-by-Step Guide to Growing Great Vegetables in a Less Than Perfect World.* Chelsea Green, White River Junction, VT.

Project BudBurst. www.windows.ucar.edu/citizen_science/budburst.

U.S. Department of Energy. "Your Home: Landscape Shading." www.energysavers.gov/your_home/landscaping/index.cfm/mytopic=11940. This web page has information about using trees to reduce energy use.

U.S. Environmental Protection Agency. "Carbon Sequestration in Agriculture and Forestry." www.epa.gov/sequestration/faq.html.

———. "Climate Change: Greenhouse Gas Emissions." www.epa.gov/climatechange/emissions/index.html. This is a good overview of greenhouse gases and includes trends and projections for U.S. greenhouse gas emissions.

———. "Improving Air Quality in Your Community." www.epa.gov/air/community/details/yardequip.html. This page has background information about two-stroke engines and air pollution, as does www.epa.gov/otaq/equip-ld.htm#consumer.

World Resources Institute. www.wri.org/climate. This website has lots of links and information about the WRI's projects and global policy efforts.

Technical Papers

Akbari, H. 2002. Shade Trees Reduce Building Energy Use and CO_2 Emissions from Power Plants. *Environmental Pollution* 116: S119–S126.

Belote, T.R., J.F. Weltzin, and R.J. Norby. 2003. Response of an Understory Plant Community to Elevated [CO_2] Depends on Differential Responses of Dominant Invasive Speices and Is Mediated by Soil Water Availability. *New Phytologist* 161: 827–835.

Bremmer, J.M., and A.M. Blackmer. 1978. Nitrous Oxide: Emission from Soils during Nitrification of Fertilizer Nitrogen. *Science* 199: 295–296.

Dukes, J.S., and H.A. Mooney. 1999. Does Global Change Increase the Success of Biological Invaders? *Trends in Ecology and the Environment* 14: 135–139.

Kern, J.S., and M.G. Johnson. 1993. Conservation Tillage Impacts on National Soil and Atmospheric Carbon Levels. *Soil Science Society of American Journal* 57: 200–217.

Lal, R. 2004. Soil Carbon Sequestration Impacts on Global Climate Change and Food Security. *Science* 304: 1623–1627.

Lenoir, J., J.C. Gégout, P.A. Marquet, P. de Ruffray, and H. Brisse. 2008. A Significant Upward Shift in Plant Species Optimum Elevation during the 20th Century. *Science* 320: 1768–1771.

Miller-Rushing, A.J., and R.B. Primack. 2008. Global Warming and Flowering Times in Thoreau's Concord. *Ecology* 89: 332–341.

North East State Foresters Association. 2002. Carbon Sequestration and Its Impacts on Forest Management in the Northeast. www.nefainfo.org/publications/carbonsequestration.pdf.

Nowak, D.J., and D.E. Crane. 2002. Carbon Storage and Sequestration by Urban Trees in the USA. *Environmental Pollution* 116: 381–389.

Pirog, R., T. Van Pelt, K. Enshayan, and E. Cook. 2001. Food, Fuel, and Freeways. Leopold Center for Sustainable Agriculture, Iowa State University. www.leopold.iastate.edu/pubs/staff/ppp/.

Qian, Y., and R. F. Follett. 2002. Assessing Soil Carbon Sequestration in Turfgrass Systems Using Long-Term Soil Testing Data. *Agronomy Journal* 94: 930–935.

Sasek, T. W., and B. R. Strain. 1988. Effects of Carbon Dioxide Enrichment on the Growth and Morphology of Kudzu *(Pueraria lobata). Weed Science* 36: 28–36.

———. 1991. Effects of CO_2 Enrichment on the Growth and Morphology of a Native and an Introduced Honeysuckle Vine. *American Journal of Botany* 78: 69–75.

8. Recycle, Reduce, Reuse, Repurpose

References for Gardeners

British Broadcasting Corporation. "UK 'landfill dustbin of Europe.'" http://news.bbc.co.uk/2/hi/7089963.stm.

Carter, M. 1997. *Garden Junk.* Penguin Books, Ltd. Middlesex, England.

Environmental Literacy Council. www.enviroliteracy.org/article.php/63.html. There are links and backgound information about landfills at this site.

Environmental Protection Agency. www.epa.gov/region07/waste/solidwaste/index.htm. This page has information about garbage.

———. www.epa.gov/solidwaste/conserve/rrr/composting/index.htm. This website has information about composting and links about composting problems in the United States.

European Environment Agency. http://ims.eionet.europa.eu/Environmental_issues/waste/indicators/generation. This page has information on the relation between economic growth and waste generation in Europe, including a link to the agency's full report.

Hawken, Paul, Amory Lovins, and L. Hunter Lovins. 1999. *Natural Capitalism: Creating the Next Industrial Revolution.* Little, Brown, and Co. New York, NY. (Also on the web at www.natcap.org.)

Kashmanian, R. M., and J. M. Keyser. 1992. "The Flip Side of Compost: What's in It, Where to Use It, and Why," in *The Environmental Gardener,* pages 15–20. Brooklyn Botanic Garden Handbook #130.

Kourik, R. 1992. "The Lazy Gardener's Guide to Recycling Yard Waste," in *The Environmental Gardener,* pages 6–14. Brooklyn Botanic Garden Handbook #130.

Royte, E. 2005. *Garbage Land.* Little, Brown, and Co. New York, NY.

INDEX

Note: page numbers in italics indicate tables. Invasive plant species listed in the Appendix do not appear in the index.